HANDY DANDY EVOLUTION REFUTER

Robert E. Kofahl, Ph.D.

Revised and expanded

Beta Books • San Diego

Handy Dandy Evolution Refuter

Beta Books, P.O. Box 23605
San Diego, California 92123

Printed in the United States of America.

Library of Congress Cataloging in Publication Data

Kofahl, Robert E
Handy-dandy evolution refuter.

Includes bibliographical references and index.
1. Evolution. 2. Bible and evolution. I. Title.
BS659.K63 575.01'62 76-54815
ISBN 0-89293-040-3

PREFACE

When Bob Kofahl and I first discussed this project, I knew we were on to something.

"Kelly, we've got to put together a book for students and parents that answers the major issues concerning evolution. A single volume with references would be ideal."

"Great, I said, "Why don't we put together a Handy-Dandy Evolution Refuter." And so Bob smiled, put on his thinking cap, and did just that.

The results will startle and amaze you. For in this volume he has compiled the major answers to the evolution controversy. Each question has its own answer with full references and a guide to additional material on that particular subject.

The answers are given in an easy to understand manner, with a simplified answer for most questions and a more detailed discussion on some of the major issues.

All in all, it's an ideal reference guide and tool for the new sport of evolution refuting. Although you may not wish to read it straight through as a book, you will find the answers you need at your fingertips.

The answers are excellent and will really prove helpful to you as you discuss the beginning of all that is. Written from the creationist's viewpoint, this book will go a long way toward refuting the *theory* of evolution.

After all, it is a Handy-Dandy Evolution Refuter.

Kelly L. Segraves
Director
Creation-Science Research Center

TABLE OF CONTENTS

Note: At the close of most sections of this expanded edition of *Handy Dandy Evolution Refuter* are selected quotations from evolutionary scientists which sustain the creationist interpretation of the data of the sciences.

Section A

SCIENCE, RELIGION, CREATION, AND EVOLUTION

1. What is science?

Answer: Science is a human activity in which people look carefully at things in the world to learn more about them. New information is discovered by means of careful observation and controlled experiments. Scientists also imagine theories to explain what they observe. Then they make more observations and experiments to test their theories. If there is no way to test a theory and possibly prove it false, it is not a theory of science. It is really philosophy or religion.[1]

2. Does science discover absolute knowledge and thus lead to absolute truth?

Answer: Science cannot discover absolute truth because science is always changing. Any scientific finding or theory may be discarded or revised tomorrow or a hundred years from now. All scientific observations and theories must be open to criticism and to possible correction or disproof. No scientific theory should be protected from criticism, because it may some day be proved to be wrong.

3. Do not scientists approach their work with a completely open mind, totally free of preconceived ideas?

Answer: All human activities are influenced by assumptions or opinions, and this includes scientific research. Every scientist has in mind some expectations or theories when he begins a research project. The new observations will tend to falsify (prove false) or to corroborate (support) the ideas he had in mind.

4. Is not evolution a science and therefore based upon fact, whereas creation is religious and therefore based upon blind faith?

Answer: Neither evolution nor creation can be tested as a scientific theory, so believers in evolution or creation must accept either view by faith.

The idea behind evolution is materialism, a belief which is accepted by faith. Materialism is the view of the world which sees matter and the laws of physics and chemistry as the only ultimate reality or, at least, the only reality with any practical importance in the world. A scientist who is a materialist will naturally pursue research aimed at discovering natural, i.e., materialistic explanations for the origin (the beginning) and development of all things including the universe, solar system, earth, life, species, man.

On the other hand, the belief underlying creation is theism, especially Biblical theism. Theism is the world-view that sees infinite-personal Spirit as the source of all reality. The Biblical theist believes that the infinite-personal Creator of all things has revealed truth about the creation in a book, the Bible. A scientist who is a Biblical theist will naturally approach scientific research with a view to understanding more about the Creator's handiwork. He will expect his results to agree with his faith, to fit into the framework of the Biblical record of creation.

It is clear, then, that evolution and creation are equally religious. Each requires faith in a basically philosophical or religious understanding of the world.

5. Can any theory about supposed processes of origins — for example, the beginning of life or the origin of the solar system — be tested in a way that might conclusively falsify (disprove) the theory?

Answer: No theory about the beginning of the world or the origin of life can be tested experimentally. This is because nobody observed or can repeat what happened in the ancient past history of the earth.

Since no humans were present at the beginning, no scientists could be on hand to record the conditions and the events. Furthermore, those conditions and events cannot be repeated experimentally. Therefore, the only evidence available is that found in the present world — in the rocks, fossils and living things. The data collected by observation and experiment in the present world, and advanced in support of one or the other theory of origins, is circumstantial evidence.[2] By circumstantial we mean that the meaning or interpretation given to the data depends strongly on the assumptions (presuppositions) of the interpreter. Furthermore, any objection raised against a theory of origins can be answered by some additional new assumption. And this new assumption cannot be tested experimentally either. Thus theories of origins, be they evolutionary or creationist, cannot conclusively be proven false by experimental test.[3] Therefore, they are really outside the realm of science. This is not to say that much true science is not involved in collecting the supporting data, but the theories are not really scientific theories.

Each person must weigh the evidence offered in support of evolution and of creation, deciding for himself on which side the evidence is most persuasive. But the tendency is to feel most heavily the evidence which is on the side of ones personal beliefs. Evolution and creation are equally scientific and equally religious. Faith is involved in the acceptance of either view, and scientific data is advanced in support of both.

6. But could not God have created everything by some evolutionary process? Perhaps He created the universe and then let evolution do the rest.

Answer: This idea, called theistic evolution, does not agree with what the Bible says about creation, and it is not acceptable to the scientists who have developed the modern theory of evolution, although it does satisfy some professing Christians.

The theory of evolution is totally materialistic, proposing evolution by completely random (chance, unplanned) chemical-physical processes. In this theory no trace of purpose, intelligence, or design is allowed. Thus theistic evolution leaves the believer with a God who really did not do what the Bible says He did, a God who apparently is not able to do what the Bible says He did. It makes man a half-evolved, half-created being who is a remodeled ape, so to speak. It also makes the Lord Jesus Christ into a very specially made-over ape. But the Bible says that He is the Creator of the universe, and the New Testament records His approval of the Genesis record of creation.[4] Thus it would appear that those who accept theistic evolution are standing in an intellectual and religious no-man's land where they will be shot at from both sides of the battlefield, by the materialists as well as by the Bible-believers.

7. I thought Darwin proved the theory of evolution in his book, *The Origin of Species*.

Answer: No scientific theory can be "proved," and the theory of evolution cannot even be tested as can the theories of experimental science. Moreover, Darwin did not use good logic in his famous book.

In 1956 W. R. Thompson, a Canadian entomologist (entomology - study of insects) of international repute, wrote in his introduction to the centennial edition of Darwin's *Origin,* "Darwin did not show in the *Origin* that species had originated by natural selection; he merely showed, on the basis of certain facts and assumptions, how this might have happened, and as he had convinced himself he was able to convince others."[5]

Chapter IV of the *Origin,* entitled "Natural Selection; or the Survival of the Fittest," occupies 44 pages in the 1958 Mentor edition. In this chapter Darwin used the language of speculation, imagination, and assumption at least 187 times. For example, pages 118 and 119 contain the following phrases: "may have been," "is supposed to," "perhaps," "If we suppose," "may still be," "we have only to suppose," "as I believe," "it is probable," "I have assumed," "are supposed," "will generally tend," "may," "will generally tend," "If," "If...assumed," "supposed," "supposed," "probably," "It seems, therefore, extremely probable," and "We may suppose." Is this really the language of science? No, it is not.

Of Darwin's speculative arguments Thompson wrote, "...Personal convictions, simple possibilities, are presented as if they were proofs, or at least valid arguments in favor of the theory... The demonstration can be modified without difficulty to fit any conceivable case. It is without scientific value, since it cannot be verified; but since the imagination has free rein, it is easy to convey the impression that a concrete example of real transmutation (change of one species to another) has been given.[5]

8. Has Christianity hindered the progress of science? Some people think so.

Answer: True biblical Christianity alone gives purpose and meaning to scientific research. Bible-believing Christians have made outstanding contributions to the progress of scientific knowledge.

The Biblical Christian faith provided the philosophical foundation for the structure of modern science. This foundation is the concept of an orderly, rational, reproducible universe purposefully designed, created, and sustained by the infinite, eternal Creator. Many of the greatest names in the history of science, men who laid the groundwork in theory and experiment for modern science, were devout believers in Jesus

Christ and in the Bible. Notable in the last century were Michael Faraday, James Clerk Maxwell, and Lord Kelvin (William Thompson). Faraday, who pioneered in experimental electricity and magnetism, is called the greatest experimental scientist of all time.[6] Maxwell developed the famous Maxwell equations for electromagnetic waves (such as radio waves) before such waves were even discovered.[7] Lord Kelvin, a great British physicist, made important contributions to thermodynamics, geophysics, and many other fields of science.[8] Each of these men was a Bible-believing Christian and a man of sterling character as well as a great man of science. Every field of science today owes much to the efforts of these and other Christians who have labored in scientific research. Furthermore, many scientists today are Christians.[9] How sad it is that so many people think science must be tied to atheistic materialism.

Popper, Karl, *Conjectures and Refutations* (New York: Basic Books, 1963), p. 257. . . . There will be well-testable theories, hardly testable theories, and non-testable theories. Those which are non-testable are of no interest to empirical scientists. They may be described as metaphysical.

Popper, Karl, *Unended Quest* (Glasgow: Fontana, Collins, 1976), p. 151. . . . I have come to the conclusion that Darwinism is not a testable scientific theory, but a metaphysical research programme —a possible framework for testable scientific theories.

Section B

THE FAILURE OF DARWIN: DESIGN IN NATURE

1. What is the theory of organic evolution?

Answer: Evolution is the theory that one or a few simple, single-celled organisms gradually changed to give rise to all of the many complex species that have ever existed. This process supposedly took several billion years.

According to this theory, several billion years ago very simple, single-celled organisms appeared on the earth. They reproduced themselves, each new generation the same as the preceding, except that in some of the new individuals random or chance changes called "mutations" occurred. Most of the mutations were bad and the individuals having them died out, but a very few of the mutations conferred some advantage on the individuals possessing them. These individuals were better adapted (adjusted) to their environment (surroundings) and so were able to reproduce more of their kind. Thus individuals having the good (advantageous) mutations gradually came to dominate the population. The favoring of a certain type of organism by the environment (nature) is called "natural selection."

According to the theory, by this stepwise process of mutation and natural selection the few original simple life forms were able to evolve (change) to more complex kinds, better adapted to their environments, to changes in their environments, or to other environments nearby. By this process ever more complex creatures supposedly originated. Many new kinds evolved to fit into new parts of the world, such as the oceans, the soil, ponds, and on the surface of and even inside of other creatures. Thus, through many millions of years new kinds of organisms evolved and flourished. Then they were replaced by new kinds, becoming extinct and often leaving their fossil remains in the rocks.

The supposed history of evolution began with single-celled animals, followed in order by single-celled plants, invertebrate (no backbone) animals, vertebrate (backboned) fish, amphibians, reptiles, mammals and birds, primates (monkeys, apes, etc.), and finally, man. So man is, according to the theory, simply the most advanced of the animals, and really only a highly organized form of matter.[10]

2. Does the theory of evolution by random (chance, unplanned) mutations (changes) and natural selection explain where all living things came from?

Answer: Many kinds of living plants and animals cannot be explained by the theory of evolution. The only other explanation is that they were created. The world is filled with these evidences that all things are the result of intelligent, purposeful design and creation by an infinite, all-powerful Creator.

A scientific theory should take into account all of the observed facts, for example, all of the species of plants and animals living on earth today. So if the theory of evolution is correct, it should be possible to imagine how a series of slight mutations could produce any existing species. And it should be possible to show how each successive mutation would have an advantage over the previous one. Finally, it should be possible to outline such a series of steps leading to each living creature and every bodily structure or organ. Such imagined histories would *not* prove evolution to be correct, but they would at least make the theory seem more plausible (likely, reasonable).

The fact is that many species and many organs completely baffle the evolutionists. In other words, the theory of evolution fails to explain the observed facts. A more reasonable explanation is that all living things were designed by an intelligent, purposeful Creator.

Let us consider a few examples of living creatures which bear silent but eloquent testimony to the reality of the infinite-personal Creator.

a. Sea Slugs With Borrowed Spear Guns.[11]

The nudibranch or sea slug, *Eolidoidea,* about two inches long, lives in the shallow tidal zone along the sea coast. It feeds primarily on sea anemones. On the tentacles of the anemones are thousands of tiny stinging cells. They explode at the slightest touch, plunging poisoned whips into intruding fish or other creatures. These are paralyzed and drawn into the anemone's stomach to be digested. But *Eolidoidea* can violently tear apart, chew up, and digest anemones without being stung, exploding the stinging cells, or digesting them. What happens to the anemone's stinging cells is one of the most amazing mysteries in nature. Connecting the sea slug's stomach with pouches on its outer surface are tiny tubes lined with cilia (moving hairs). The cilia sweep the undigested stinging cells from the stomach to the pouches where they are arranged and stored for the sea slug's defence. Thus when a hungry fish tries to take a bite of sea slug, it gets stung in the mouth by the stinging cells which the hapless anemone manufactured for its own defense!

Stop a moment and try to imagine the complicated series of minute changes or adaptations which would have to take place in both the sea slug and the anemone in order to produce this relationship. The slug would have to evolve some chemical means of preventing the stinging cells from exploding, even with the roughest handling. The anemone would have to cooperate by refraining from evolving countermeasures, even though according to the theory of evolution it would be expected to evolve defenses against every threat in its environment. And the sea slug would have to evolve an altered digestive system which does not digest stinging cells but does digest the other tissues of the anemone. The sea slug would also have to evolve the ciliated tubes leading to the pouches which it was also evolving on its outer surface. Finally the slug would have to evolve subtle mechanisms for arranging and storing the stinging cells and keeping them alive in the pouches. If the evolutionary theory is to succeed in explaining these developments, it must be able to show how a multitude of slight changes, each one advantageous to the sea slug or to the anemone, could produce the finished relationship. And, of course, the theory should also explain why the sea anemone would put up with this outrage. After all, evolution is supposed to be good for anemones, too.

No suitable explanation for the evolution of this mystery has surfaced to date. Can the community of evolutionary scientists come up with an explanation? Their failure to do so thus far is a tacit admission that creation is the only reasonable answer. These creatures obviously were designed and created to fit into their relationship.

b. Ants in the Plants

The Bull's Horn acacia tree of Central and South America is furnished with large hollow thorns. A species of ferocious stinging ants pierces the thorns to use as nests. Small green bumps on the twigs and little brown nubbins on the leaf tips serve as food for the ants. Thus the ants get food and shelter from the tree, but what does the tree get out of the deal? It gains total protection. No goat, not even a leaf hopper, is going to chew on this tree with impunity, for the ants attack any intruder. But there is something even more amazing about the symbiotic (living together) relationship between the tree and the ants. These ants are gardeners! As is well known, in a tropical jungle the competition for space and sunlight is very intense. Some vines specialize in strangling other plants. But the Bull's Horn acacia has no problem. The ants make regular forays in all directions from their home tree and nip off every green shoot that threatens to encroach on their tree house. So this tree always has plenty of space and sunlight. In actual experiments when all of the ants are removed from one of these trees, the tree dies within two to fifteen months, eaten by foraging animals and insects or suffocated by the surrounding jungle.[12]

Our question for the evolutionary theorists is this: who taught the ants to be gardeners? Odum in his excellent book, *Fundamentals of Ecology,* says that this is an example of "co-adaptation," but he does not venture to explain the sequence of minute coordinated changes or adaptations in the tree and the ants by which the alleged historical process of co-adaptation occurred.[12] Thus the evolutionary explanation remains a faith, not a science. Does not the creationist faith appear more rational?

c. Birds, Stars, Maps, and Compasses

Migrating birds perform amazing feats of navigation which are still not fully understood by scientists. The lesser whitethroated warbler summers in Germany but winters near the headwaters of the Nile River in Africa. Toward the close of summer, when the new brood of young is independent, the parent birds take off for Africa, leaving their children behind. Several weeks later the new generation birds take off and fly, unguided, across thousands of miles of unfamiliar land and sea to join their parents. And they have never been there before. How do they navigate? German researchers raised some of the warblers entirely in a planetarium building. Experiments proved that in their little bird brains is the inherited knowledge of how to tell direction, latitude, and longitude by the stars, plus a calendar and a clock, plus the necessary navigational data to enable them to fly unguided to the precise place on the globe where they can join their parents! Many other species of birds perform similar feats. How did all of this knowledge and ability get packed into a little bird egg?[13]

Recent investigations at Cornell University have revealed that the homing pigeon determines directions by observing the position of the sun in relation to the bird's internal calendar and clock. But pigeons were also shown to have direction-finding ability in cloudy weather when the sun was not visible. Electromagnets placed on the pigeons' heads destroyed this ability in cloudy weather, but not in sunny weather. Thus, the pigeon has two ways of determining direction, by the sun in clear weather and by an internal magnetic compass of some sort in cloudy weather. There is also good evidence that pigeons have an internal map of the earth which they use in conjunction with their compasses to enable them to navigate accurately over distances of hundreds of miles.[14]

Science cannot explain how these remarkable abilities of "scatterbrained" birds could have evolved by chance. A more reasonable explanation is that birds were designed and created with those abilities required for a happy life with the birds.

d. Good Engineering in the Human Body

The human body is amazing not only because of its excellently engineered structures and mechanisms such as bones, joints, muscles, communications networks, and circulatory systems, but also because of the many sophisticated control systems which regulate all of the bodily functions. In general these control systems use the principle of negative feedback which is basic to the control mechanisms designed by engineers for use in automobiles, air conditioning systems, and manufacturing plants. Physiologist (physiology - study of life processes, activities, and functions) David A. Kaufman lists ten different classes of control systems in the human body: internal environment and homeostasis (keeping internal conditions such as temperature constant), nervous control, hormone control (hormones are messenger chemicals), contractile control, circulatory control, respiratory control, electrolyte control, digestion and absorption, resting and energy metabolism (producing energy from food), and regeneration and reproduction.[15] There is not space here even to define all of these bodily functions, but one example will be described.

The human body has a temperature control system which keeps the body core temperature relatively constant at 99.6°F. The hypothalamus, a gland in the brain, contains an unknown device which provides the standard or ideal temperature signal. This signal goes to a comparison device where it is compared with the temperature signal from a temperature sensor which reads the actual body temperature. From the comparison device an error signal of too hot or too cold is sent to either the body's antirise or its antidrop center. If stimulated, the antirise center turns on the sweating and vasodilation (expanding of blood vessels) mechanisms. These release heat to the surroundings and reduce bodily heat production. On the other hand, the antidrop center, if stimulated, turns on the shivering and vasoconstriction (constriction of blood vessels) mechanisms. These preserve heat and increase bodily heat production.

Scientists only vaguely understand the details of this efficient control mechanism, which are undoubtedly highly sophisticated. What great faith in materialism is required for an informed person to believe that these intricate engineering techniques utilized in the human body and absolutely essential to human existence could be the end result of blind, purposeless evolution.

The four examples just considered illustrate the fact that the theory of evolution fails dismally when the structures and functions of living things are examined in detail. We have presented these and other such examples of intelligent, purposeful design in nature before many college and university audiences. Not once has a qualified scientist or faculty member offered to explain their origin by evolution through many intermediate steps. A theory which majors on broad generalities but fails on the details is not very satisfactory science.

Deevey, Edward, Jr., *Yale Review, 61,* Summer (1967), pp. 634-5.
. . . Of course these things are marvels, and of course, the fossil record being what it is, no one can say with confidence exactly how any one of them came about.

Section C

LIFE FROM CHEMICALS: THEORY AND IMPROBABILITY

1. Many scientists seem quite sure that in the atmosphere and oceans of the early earth large quantities of the life building-block molecules such as amino acids were produced. Is this a reasonable theory?

Answer: This theory is purely a set of guesses which cannot be proven to be true. There are many difficulties with this scheme.

a. The oxygen-ultraviolet dilemma

Their assumed atmosphere can contain no oxygen, for oxygen eats up amino acids. But without oxygen there could be no ozone layer high in the atmosphere. The ozone layer in our present atmosphere stops the sun's ultraviolet rays which would destroy amino acids and other building block molecules. So here are the horns of the oxygen-ultraviolet dilemma: 1) oxygen present — amino acids destroyed by oxygen, and 2) oxygen absent—amino acids destroyed by ultraviolet light.

To get around this difficulty more assumptions are made. Perhaps amino acids were protected in bottom waters of shallow lakes, or the ocean surface may have been covered by a layer of tar-like chemicals which stopped ultraviolet light. Thus the theory becomes a web of unprovable assumptions.[16]

b. Oxygen-free atmosphere only an assumption

The presence on the early earth of a reducing (no oxygen present) atmosphere containing methane and ammonia is only an assumption required to make the theory work. There is no conclusive evidence to justify the assumption.[17] Some of the other planets, notably Jupiter and Saturn, have reducing atmospheres, although Mars has a very thin oxidizing atmosphere.

It can be added that the sun's ultraviolet light breaks down water molecules to release free oxygen at such a rate that the ancient atmosphere could not be free of oxygen for long.[18] Ammonia also is decomposed by ultraviolet light and would soon drop to a concentration too low to participate in a chemical beginning of life.[19]

The more recent idea that cynanide produced by reactions in the earth's crust provided the raw material to make amino acids, etc., also suffers from numerous problems including the following.

c. Many unsolved chemical problems.

A great deal of research in recent years has been aimed at showing how chance chemical reactions in an ancient atmosphere and ocean could have produced the molecules needed for life. However, plausible methods of accidental synthesis on the assumed early earth have not yet been discovered for many of the necessary substances. These include five of the twenty required amino acids as well as fatty acids, the sugar ribose, DNA and RNA, and the building blocks called nucleotides for making DNA and RNA.[20]

d. Building block molecules too unstable.

Many of the necessary compounds mentioned above, if they could be formed, are not stable enough to be accumulated. They would decompose or react together in the wrong way before they could take part in a beginning of life experiment. This problem is serious in the cases of sugars, about half of the twenty amino acids, DNA and RNA, and several of the DNA and RNA building blocks.[21]

2. Could amino acids have linked together by chance to form the long protein chains, and could nucleotides have combined accidentally to form the long DNA and RNA molecular chains?

Answer: The tendency (because of the entropy effect — See Section D-1) is for the protein and DNA chains not to form, but to break up.[22] Nevertheless, chemists have discovered several types of reactions by which protein molecules might be produced spontaneously under certain very special conditions.[23] However, that these unlikely conditions ever existed on the earth is only a highly optimistic assumption. But even if protein-like molecules were actually to form by such random processes, the probability is vanishingly small that the right ones to start life would ever form. For a calculation of this improbability see the next question.

3. Admittedly the chance chemical beginning of life was a very improbable event. But wasn't there enough time for it to happen anyway?

Answer: Even with trillions of years there would not be enough time to make it probable that chance chemical reactions could form even the simplest living organism.

The argument that sufficient time makes anything possible or even probable sounds plausible only if it is not analyzed carefully. It starts with the admission that, since even the simplest living organisms are exceedingly complicated, the beginning of life by accidental chemical reactions is very improbable. The probability is very, very low that just the right molecules would form together and spontaneously fit together to start life. But if a very unlikely thing is tried many times, the probability increases that success will finally be achieved. If there is enough time to make a large enough number of tries. the mathematical probability that it will finally happen becomes almost certainty.

Mathematically, this argument is correct. But to see if the mathematical theory really proves that life could have started accidentally, it is necessary to apply the theory to a reasonable model of the real world. We do this in the Appendix to this book and in fuller detail in our book, *The Creation Explanation.* [24] We begin with very generous assumptions about the beginning of life. Then, we assume that for a billion years the surface of the earth was covered each year with a fresh layer one foot deep of protein molecules. This would be 260 trillion tons each year, a fantastic number of molecules. Yet, at the end of the billion years, the probability that just one protein molecule required to start life had been formed is only one chance in about 100 billion. This means that it is really mathematically impossible for life to start by accident, even if the beginning would require only a single suitable enzyme molecule.

Some workers have claimed evidence that certain origin-of-life experiments have produced chains of amino acids which were non-random in order. Supposedly certain sequences of amino acids tend to form, and reportedly these sequences are similar to those found in true proteins.[25] On the other hand, Miller and Orgel challenge such claims and say, "There is no evidence to show whether the amino acids within a chain are highly ordered or not."[26]

In any event it is quite certain that life could not start with a single protein molecule. It has been estimated by Harold Morowitz that the simplest possible living cell would require not just one, but at least 124 different proteins to carry out necessary life functions.[27] Writing in his book, *Energy Flow in Biology,* Professor Morowitz also estimates the probability for the chance formation of the smallest, simplest form of living organism known today.[28] He comes up with the unimaginably small probability of one chance in $10^{340\,000\,000}$ This means one chance in the number formed by one followed by 340 million zeros. This is about the same as the probability of tossing a coin one billion times and getting all heads! Nevertheless, Dr. Morowitz and thousands of other capable scientists believe that life happened by accident. But must you and I accept their unreasonable faith in materialism? Is not the Biblical faith in the all-powerful, all-knowing, infinite-personal Creator the better, the more reasonable faith?

4. Could the genetic code have originated by chance?

Answer: Scientists who believe it did, and there are many of them, have failed to find a plausible explanation of how it did.

The genetic code is the code by which the long DNA chain molecules carry the instructions for arranging the amino acids in the proper order along the long protein chain molecules. Four different link-molecules called nucleotides make up the DNA chain. These are referred to by their abbreviations, A, C, G, and T. A group of any three of these "letter molecules" in a DNA chain forms a code word for one of the twenty kinds of amino acids which make up the protein chains in living organisms. For example, the DNA chain, CTA, is the code word for the amino acid leucine.

A gene consists of a long chain of the DNA code words corresponding to a long chain of amino acids to be linked together to form a particular protein molecule. How is the meaning of the code message translated into the protein molecule? It is a very complex process only partially understood by scientists. The DNA code message or gene is transcribed into a similar code message on a messenger RNA (m-RNA) molecule. The message on the m-RNA molecule is read by a ribosome (a very complex object made of some 55 different protein molecules and a roughly equal weight of long RNA molecules). The required amino acid molecules are recognized by transport RNA (t-RNA) molecules. Each type of amino acid is attached to its own special type of t-RNA molecule which transports it to the ribosome which then connects the amino acid molecule onto the growing protein according to the code message in the m-RNA molecule. All of these various steps are made possible by the assistance of various enzyme molecules.

Is it not amazing that all of this complex process supposedly arose without any Designer, without any purpose? Miller and Orgel admit in their very honest book, "We clearly do not understand how the code originated. New ideas that can be tested experimentally are needed."[29]

Miller, Stanley L., and Orgel, Leslie E., *The Origins of Life on the Earth* (Englewood Cliffs, N.J.: Prentice-Hall, Inc., 1974), p. 33.

Geological and geophysical evidence is insufficient to allow us to state with any precision what conditions were like on the surface of the primitive earth. Arguments concerning the composition of the primitive atmosphere are particularly controversial. It is important, therefore, to state our own prejudice clearly. We believe that there must have been a period when the earth's atmosphere was reducing, because the synthesis of compounds of biological interest takes place only under reducing conditions.

Yockey, H. P., "A Calculation of the Probability of Spontaneous Biogenesis by Information Theory," *J. Theor. Biology* (1977) *67*, pp. 393, 396.

With regard to the appearance of a single molecule of the cytochrome *c* family, even the *deus ex machina* needs 10^{36} "acceptable planets" with just the right conditions for 10^9 years. . . One who finds the chance appearance of cytochrome *c* a credible event must have the faith of Job. . . .

. . . The "warm little pond" scenario was invented *ad hoc* to serve as a materialistic reductionist explanation of the origin of life. It is unsupported by any other evidence and it will remain *ad hoc* until such evidence is found. Even if it existed, as described in the scenario, it nevertheless falls very far short indeed of achieving the purpose of its authors even with the aid of a *deus ex machina*. One must conclude that, contrary to the established and current wisdom a scenario describing the genesis of life on earth by chance and natural causes which can be accepted on the basis of fact and not faith has not yet been written.

Section D

WOULD EVOLUTION VIOLATE KNOWN PHYSICAL LAW?

1. Is evolution contrary to the law of increasing entropy (increasing disorder)?

Answer: One of the most firmly established laws of physics (it could be called The Natural Law of Degeneration) states that all natural processes in the universe cause a net increase in disorder and a net loss of useable energy.[30] This experimentally established law is contradicted by the theory of evolution, according to which natural processes taking billions of years changed disordered and simple molecules into the complex, highly ordered, energy-rich structures of living things. A burning candle can serve to illustrate the Natural Law of Degeneration.

The paraffin wax in a candle is composed of hydrocarbon molecules. These are long chains of carbon atoms to which are attached hydrogen molecules. Thus, paraffin molecules are structured; also, they possess much chemical energy. This is energy in a form that can do work. If a candle is lit, it will burn spontaneously (naturally, automatically). Oxygen from the air will combine with the carbon and hydrogen to produce water vapor and carbon dioxide gas. The chemical energy will be converted into light and heat energy. The candle burns spontaneously, but never will a candle "unburn" itself and reform itself spontaneously. The second law of thermodynamics describes this uniform behavior of the natural world as follows: all spontaneous processes change complexity into disorder and organized energy into random heat energy. Thus, order (the form of the candle and the structure of the large paraffin molecules) was transformed into disorder (puddles of unburnt wax and the much smaller molecules of carbon dioxide gas and water vapor scattered throughout the air in the room). Also, organized energy (the chemical energy packed into the paraffin molecules) was changed into heat energy in the air, walls, and objects in the room. This energy is now in a form that is less available to do work.

The increase in disorder or randomness of the structure and energy of a physical system is measured as an increase of a property of the system called "entropy." So the second law predicts that entropy will increase. But according to the theory of evolution entropy spontaneously decreased. Thus evolution is supposed to be a natural process which occurred according to the laws of physics, yet transformed completely disordered matter into highly ordered, energy-rich living organisms. And supposedly these organisms continually increased in order and complexity by evolution. Surely in this there is an enigma: scientists who accept the second law of thermodynamics as valid believe nevertheless that a natural process called evolution, which violates that law of physics by causing entropy to decrease, has continued for billions of years.

2. Do evolutionary scientists offer a solution to this difficulty with the second law, the law of increasing entropy, of increasing disorder?

Answer: The standard answer to the entropy problem is that the flow of energy from the sun could reverse the processes of degeneration locally on the earth's surface without violating the entropy law in the earth-sun system taken as a whole.

This argument is based upon the fact that the earth where life supposedly evolved is an open system, not a closed system. That is, energy from the sun continually flows through the earth's atmosphere and to its surface. Therefore, the earth and the sun must be considered together. There can be a large increase in entropy in the sun at the same time a small decrease occurs on earth. Supposedly, then, life could begin by chance on the earth, with a modest decrease in entropy, increase in complexity. At the same time the sun would have expended a huge amount of energy, with a very large increase in entropy. So the net result in the earth-sun system would be an increase in entropy, and the second law would be satisfied in this system.

Let us develop this idea further to see if it is valid. It is a fact that when some chemical mixtures are irradiated with light, energy is absorbed and new reactions occur.[31] Some molecules are produced which are more complex and energy-rich than those in the original mixture. The argument is that if this can happen to a small extent in the laboratory, a similar process continued for billions of years on the earth's surface could have initiated life. It supposedly could have formed the entire biosphere out of non-living chemicals, without violating the entropy law for the earth-sun system.

3. Is this open-system argument really valid? Can it stand critical analysis?

Answer: Local reversals of natural degeneration (i.e., of entropy increase) can be only very limited and temporary. The open-system argument cannot be shown to be valid unless scientists can show experimentally that chance or random chemical reactions can actually bring about unlimited increase in chemical complexity and produce something having the properties of life. All chemical research on the origin of life has thus far failed at both of these points.

First, when light is absorbed by a chemical mixture, any formation of molecules with more complicated structures and more energy content soon reaches a limit. This limit is far short of the complexity of the simplest living organism. The reason for this is that energy-rich molecules tend to break down. Their free energy (energy that can do work) tends to be changed into heat. Therefore, the build-up of complexity and energy content reaches a dynamic equilibrium point and stops. Pushing the process further by pumping in more radiation or higher energy (shorter wave length) light photons would soon start breaking up molecules and overheating the mixture. Only limited increase in chemical complexity has ever been achieved in such experiments.

Second, to capture light energy and use it to build up the structures in living things requires the complex machinery already present in those organisms. The photosynthetic apparatus of green plants is the prime example. Without such energy-capturing and energy-using systems, the effect of light energy is mainly to break down complex structures. An example of this is sunburn, the destruction of skin tissue by the sun's ultraviolet light.

Third, no laboratory experiments have demonstrated the production of coded information, reproduction, or other essential attributes of living systems. By random chemical reactions no self-contained structure has been produced which exhibits the unique set of eight properties characteristic of life. These are (1) a stable, complex structure, (2) metabolism (use of materials and energy from outside to build and operate the system), (3) growth, (4) homeostasis (maintaining constant internal conditions, e.g., body temperature or chemical concentrations), (5) response to environment, (6) reproduction, (7) adaptability and (8) coded information.

Until it can be demonstrated that some mixture of non-living chemicals can originate structures with these eight characteristics of life, the second law remains a forceful refutation of evolutionary theories.

4. Does the entropy law have anything to do with mutations? If so, can mutations be expected to produce increasingly complex order in living creatures?

Answer: Since mutations occur according to the law of increasing entropy (disorder), it is not reasonable to believe that they could produce increasing order. The majority of mutations appear to be bad, destroying order.

Each time cell division occurs the DNA molecules of the genes must be copied so that the resulting daughter cells have the same coded information as the parent cell. A large proportion of the mutations in wild populations probably result from errors in the copying process. These may well be caused by the random heat motions of the molecules when the DNA is being copied.[32] Radiation, certain chemicals, and other influences can also cause mutations. In accordance with the second law the effect of mutations should be to make the sequence of the letters of the genetic code message carried by the DNA molecules become more disordered or random. That is, the coded message or information in the DNA molecules should tend to become on the average less and less meaningful, more and more scrambled, until it becomes nonsense after many cell divisions. Therefore, mutations should not be expected to produce increasingly complex and meaningful information content in the DNA of any species. Rather, the result should be just the opposite. This is verified in nature where mutations appear to be bad for any organism.[33]

5. But does not natural selection solve this problem and reverse the second law of thermodynamics by filtering out bad mutations and preserving good ones?

Answer: It has not yet been shown experimentally that mutations and natural selection can produce new structures or new organs. There is no question that the pressure of the environment acts to remove from the population those types which are not so well adapted. Thus natural selection is primarily conservative, preserving the normal or wild type.[34] But the vital question is whether or not mutations can be preserved and accumulated by natural selection so as to produce creatures with new structures, new organs, new behavioral patterns, etc. This is the essential point which has yet to be demonstrated by observation or controlled experiment. The belief that evolution *can* happen and that it has happened in the past is more a faith than a scientific theory.[1,2,3]

6. Do living cells violate the second law?

Answer: Living cells do not violate the second law. They merely overcome its effects for a limited time.

Living cells, while not violating the entropy law, do overcome it for a while by feeding on free energy(energy available to do work) in food taken from the surroundings.[35] Non-living systems cannot do this. The chemical structures in living cells are held together only by fairly weak bonding forces. Therefore, they are easily broken down by the random heat motions of the molecules. This is the entropy effect. In a dead cell this process soon reduces the cell to formless rubbish. In a living cell energy imported from outside powers a multitude of repair projects which operate continually. They are guided by the plans carried in the DNA molecule, so that the correct cell structure is preserved. The DNA molecule is held together by stronger forces than is the general cell structure; therefore the plans are not normally altered by the molecular motions. Gradually, however, the degenerative entropy effect does cause the breakdown of some of the more permanent structures of the cell, including perhaps alteration of the DNA, and leads to aging and death. Non-living systems cannot duplicate the entropy-overcoming processes of living cells.

Jacobson, Homer, *Amer. Scientist, 43,* Jan. 1955, p. 121.

Life, then, is a temporary reversal of a universal trend by means of the production of information mechanisms. . . . All of these processes have in common the production of information. This minor reversal of entropy gain is probably an accident in the history of the universe. The reproductive facility of the organisms *maintains* the phenomenon. The accident—if it was an accident—presumably happened at some early time, or times, corresponding to the first coalescence of matter into a self-reproducing structure which could extract energy from the environment for its first self-assembly. Directions for the reproduction of plans, for energy and the extraction of parts from the current environment, for the growth sequence, and for the effector mechanisms translating instructions into growth—*all* had to be simultaneously present at that moment. This combination of events has seemed an incredibly unlikely happenstance, and has often been ascribed to divine intervention.

Section E

CAN MUTATIONS CREATE NEW SPECIES?

1. Isn't it possible that slight changes produced by mutations could add up over many generations? Doesn't this explain how evolution could form new structures and organs and even entirely new kinds of plants and animals over long periods of time?

Answer: The vast majority of mutations are admitted by the evolutionists themselves to be bad. Furthermore, it has not been shown experimentally that mutations and natural selection can produce new organs. Limited changes have been observed in many species, but that does not prove that these species could evolve in millions of years into entirely different kinds of creatures. Scientists merely assume that this happened in the past when no scientists were present to observe the process.

A mutation is a random change in a gene. Every individual organism inherits a set of genes from its parent or parents. Apparently most genes contain coded instructions for building the thousands of different protein molecules found in living cells. The average gene contains 600 to 1800 precisely ordered code letters. A mutation which changes, adds, or subtracts a single letter can change the coded message and thus modify the resulting protein. A very slight change — in fact most changes in a protein molecule — can cause it to function poorly or not at all. As a result the organism usually is not as viable (able to live) as the wild strain of the organism before mutation, and many mutations are lethal (deadly). Geneticists have concluded that the vast majority of mutations are bad. Sir Julian Huxley estimated that perhaps less than one-tenth percent of all mutations could be advantageous to an organism.[33] This cannot be quantitatively demonstrated by experiment however. Of the remainder some are apparently neutral, but the large majority either weaken or kill the individual.[36]

The pressures of the environment and the necessities of life tend to eliminate from the population those mutations which lower the ability of the organism to reproduce itself. This effect is called "natural selection." Thus natural selection is seen to be a conservative process which tends to preserve the normal wild type in the population and to eliminate most innovations (new things).

According to the current theory, however, the tiny percentage of mutations which are neutral or helpful provide the new design information, this, when added to many other mutations occurring in the course of many generations, results in a population better adapted to the environment. This supposedly leads to new structures and organs and even to entirely new creatures. So, for example, reptiles supposedly evolved their scales into feathers and changed into birds.

2. Are there many difficulties with evolution by mutation and natural selection as a scientific theory?

Answer: Yes, there are many difficulties with the mutation-natural selection theory of evolution, for example:

a. History can't be proved by science.

The alleged historical process of evolution from amoeba to man was not observed by man and cannot be reproduced experimentally. Thus the claimed fossil evidence is only circumstantial (See A-5). That two fossils of extinct creatures are genetically related through a long series of mutations in many generations is an unverifiable assumption in every case. It is even difficult to demonstrate that two living similar species are genetically related.

b. Structures too complex.

The genetic structure, the proteins coded by the genes, and the bodily structures, and the metabolism of all organisms are exceedingly complex and very delicately balanced systems. The theory that such systems could have been produced and can be improved by mutations is like saying that an electronic computer could be produced and then improved by shooting bullets into a room stuffed full of computer parts.[37]

c. Wild types stronger than mutated types.

A very few experimentally observed mutations in the fruit fly, *Drosophila,* reportedly confer a slight advantage under special conditions in the laboratory.[38] However, the observed mutated flies have proven generally inferior to the wild type, and under ordinary conditions populations tend to revert to the wild type. The DDT-resistant populations of houseflies illustrate this fact. They do not reproduce as effectively as the wild type in the absence of DDT.

d. Change limited in microorganisms.

In the case of microorganisms, many mutations have been observed which reportedly confer advantage in specialized environments—for example, in the presence of antibotics. Nevertheless, new species have not been produced, only new strains. Under normal conditions such populations tend to revert back to the original wild type, and bacterial populations tend to have great genetic stability.[39]

e. Mutations only modify what already exists.

Mutation and natural selection can only modify that which already exists, for example, changing size, color, efficiency of operation, etc. The tendency is to preserve, not to innovate.[34] There is no evidence that a mutation or series of mutations has ever created a new structure or organ.

f. Theorists still arguing.

The actual mechanism of supposed evolution is still a matter of debate among evolutionary theorists. A current question concerns whether evolution progresses mainly by natural selection of advantageous mutations or by the accumulation of neutral mutations.[40] If after a hundred years the specialists can't agree on the essential mechanism of their theory, perhaps the whole idea is wrong.

g. Conclusion: mutations inadequate.

We would conclude that the production of new design information by gene mutations is entirely inadequate to explain evolution, and that mutations and natural selection can bring about only limited variation of already existing designs. The majority of mutations seem to be bad, not creative. The accumulation of mutations seems to be limited, leading to destruction, not to new design.

h. Chromosome changes also inadequate.

The other types of genetic change, such as the translocations and inversions of parts of chromosomes, only rearrange already existing design information. Such processes cannot explain the origin of all the different kinds of living things.[41]

i. Source of new genes not established.

A suitable mechanism for new genes to be formed has yet to be demonstrated with any certainty. One popular theory is that a duplicate or unused gene — sometimes termed a "floating gene" — can be "preadapted" by random neutral mutations until it becomes useful for another purpose.[42] While this idea may sound plausible, when all of the necessary details are taken into consideration it is rather unconvincing. In cases in which such a gene transformation has been reported, it is possible that the changed gene was a damaged one which was repaired by a series of reverse mutations.

3. Are there not many examples today of plants and animals which have changed and thereby shown evolution in action?

Answer: All of the examples are far too limited to explain evolution "from amoeba to man."

There is no question that change in populations occurs, but those changes which have been observed are very minor compared to what is needed to make evolution possible.[43] The changes actually observed are merely variations in already existing structures. The uniform testimony from genetics indicates only limited change, and that absolute boundaries exist between different kinds of organisms. This is consistent with the Biblical record of the kinds created to reproduce each after its kind. The Bible does not precisely define the boundaries of the created kinds, and thus this question is a basis for scientific research. As an example, all the Canidae (dogs, wolves, coyotes, foxes, and jackals) are apparently capable of interbreeding. They must, therefore, belong to the same original kind. Likewise, the various cattle, buffalos, and bison also interbreed, so they must belong to the same kind. But as is well known, there is no interbreeding between dogs and cattle. They are certainly different Biblical kinds. In the original creation several different types, having the potential to interbreed, may have been created within an original kind.

4. Is not Kettlewell's moth in England a proven example of evolution in action by means of mutation and natural selection?

Answer: The change observed in Kettlewell's moth produces a different phase, not a new species. It is not evolution.

Before air pollution associated with the industrial revolution darkly stained the tree bark and killed the light colored lichens on the trees, the moth population was primarily light colored. However, the dark phase apparently existed in the population also.[44] As the trees gradually grew darker, birds could better see to pick off the light colored moths, so the population became dominated by the dark colored moths.

Nevertheless, the species, *Biston betularia,* remains the same.[45] And now that the English air is being cleaned up, it is reported that the proportion of light colored moths is again on the increase. The case of Kettlewell's moths may demonstrate natural selection in action, but not evolution of a new kind of creature. This is evidence for amoeba-to-man evolution only to one who already believes. And that is what evolution is, a belief, just as much as is creation.

Goldschmidt, Richard B., *Amer. Scientist, 40* (1952), pp. 94, 97. . . . It is true that nobody thus far has produced a new species or genus, etc., by macromutations. It is equally true that nobody has produced even a species by the selection of micromutations. . . . Neither has anyone witnessed the production of a new specimen of a higher taxonomic category by selection of micromutants.

Wald, George, *Math. Challenges to the Darwinian Interpretation of Evolution*, Moorehead & Kaplan, eds. (Philadelphia: Wistar Inst. Press, 1967), pp. 18, 19. . . . I took a little trouble to find whether a single animo acid change in a hemoglobin mutation is known that doesn't affect seriously the function of that hemoglobin. One is hard put to find such an instance. . . . The restrictions are enormous.

Section F

FOSSILS — CREATED OR EVOLVED?

1. Do fossils illustrate a gradual, step-by-step change of life from very simple to more complex creatures, from primitive invertebrates to vertebrates to man?

Answer: The fossil record has many serious gaps, so that the intermediate forms needed to connect different kinds of creatures are missing. This information from the fossils fits the idea of creation rather than of evolution.

The accompanying chart shows the so-called geologic column with the different geological eras and periods and some of the types of fossils. The order shown from bottom to top is that in which they supposedly evolved. This is the order in which they are expected normally to be found in the rocks if evolution is a correct theory. This chart looks neat and seems to agree with the evolution theory, but when one looks at all of the facts, the picture is not really that convincing.

2. Does the fossil record show gradual evolution from single cells to simple invertebrates to complex invertebrate animals?

Answer: The fossil record has many embarrassing gaps, even reversals. It has been said that the record is "composed mainly of gaps" and that it does not provide intermediate forms between species. The most striking gap is that between single-celled organisms and the complex invertebrates.

The rocks containing the supposedly oldest assemblage of marine invertebrate species are called Cambrian rocks. The fossils are supposed to represent "simple, primitive" forms. In actuality many of the Cambrian creatures are highly organized and complex, and some are almost indistinguishable from modern forms. Furthermore, in the Cambrian rocks are found representatives of every major plylum or grouping of animal life except one, the vertebrates, which are classified as a subphylum of the phylum Chordata.[46]

So the supposedly oldest rocks containing appreciable numbers of fossils include many different kinds of structurally complex creatures. But according to evolution one would expect a few very simple kinds, if any organism can really be considered simple.

3. But what of the rocks which supposedly were deposited before the Cambrian rocks, and which should be found underneath them?

UNIFORMITARIAN GEOLOGICAL TIME SCALE

Era	Period	Beginning, assumed years before present	Assumed sequence of evolving life forms
Cenozoic	Quaternary	2,000,000	Modern man, plants, animals
	Tertiary	60,000,000	Mammals dominant, modern birds
Mesozoic	Cretaceous	130,000,000	Mammals multiply, flowering plants, dinosaur extinction
	Jurassic	180,000,000	Reptiles dominant, first mammals, first toothed birds
	Triassic	230,000,000	First dinosaurs, mammal-like reptiles
Paleozoic	Permian	280,000,000	Reptiles displace amphibians, modern insects, evergreens

	Pennsylvanian	310,000,000	Reptiles from amphibians
	Mississippian	340,000,000	Winged insects, bony fish
	Devonian	400,000,000	First amphibians insects
	Silurian	450,000,000	First land animals (arthropods)
	Ordovician	500,000,000	First vertebrates (fish), land plants
	Cambrian	570,000,000	Abundant marine invertebrates, trilobites abundant
Proterozoic (Precambrian)		2,000,000,000	Algae, microorganisms
Archaeozoic (Precambrian)		4,500,000,000	Origin of life (living cells from non-living chemicals)

Answer: These rocks, classified together under the term Precambrian, contain no fossils other than some single-celled types such as bacteria and algae. The Cambrian rocks contain a wealth of complex fossils as indicated above. Where are the intermediate forms which represent the greater part of the history of evolution? They are nowhere to be found. This is perhaps the most striking and perplexing gap in the fossil record which confronts the theory of evolution.[47] For over a century paleontologists have searched for fossils to fill the gap, but without success.

One might suspect that the entire fossil record has been misinterpreted because of the materialist assumptions of the interpreters. Rather than a gradual evolution from simple to complex, the Cambrian fossils can be interpreted in terms of sudden creation of many complex types of marine creatures which lived together in the same sea bottom environment. Since they lived together, sudden dumping of sediments by the Flood trapped and fossilized them together.

4. Are the invertebrate ancestors of the vertebrates known?

Answer: In evolutionary theory some unknown Cambrian invertebrate evolved a backbone and became the first vertebrate fish. However, the fossil gap between the unknown Cambrian invertebrate ancestor and the first vertebrate fish is said to be 100 million years with no fossil evidence. Furthermore, the various orders of fish appear from the fossil record to have arrived on the scene completely separate and distinct from the beginning.[48]

5. Does fossil evidence exist for gradual evolution of fish into amphibians?

Answer: The gap between fish and amphibians is an assumed period of millions of years without necessary fossil transitional forms. There should have been a multitude of such forms leading from the fin of the crossopterygian fish to the leg of the ichthyostegid amphibian. None has been found.[49] Another difficulty stems from the forms of the vertebrae in supposed evolutionary series. Both the crossopterygians and ichthyostegids had arch type vertebrae.[49] On the other hand, the three modern fossil orders of amphibians which allegedly evolved from them had supposedly more primitive vertebrae of the so-called "husk vertebrae" type. Strangely, the three living orders of amphibians also have the "primitive" type of vertebrae. Finally, none of these groups of fish or amphibians are connected by series of intermediate types.[50, 51]

6. Is the alleged evolution of amphibians into reptiles well documented?

Answer: The gap between the amphibians and the reptiles is found in the fact that the foremost candidates for the key transition amphibians, *Seymouria* and *Diadectes,* are found in the wrong rocks. They supposedly appeared some twenty million years after the appearance of the original reptile group and also after the appearance of the reptile group from which the mammals are said to have evolved.[52] How can the parents appear after the children? Another problem is the requirement that the simple gelatinous amphibian egg, designed to develop in water, be transformed by slow, minute changes into the complex amniotic egg of the reptiles, designed to incubate in air. There is no direct fossil evidence for this transformation, and it is difficult indeed to imagine how it could have occurred.[53]

7. Is there a fossil gap between reptiles and mammals?

Answer: The gap between the reptiles and the mammals is especially striking with respect to the bones of the lower jaw and the ear bones or ossicles. All reptiles have at least four bones on each side in the lower jaw and one bone in the ear. All mammals, living or extinct, have one jaw bone and three ossicles[54]. The gradual evolution from the simple reptile ear to the delicate, precisely engineered human ear, for example, is impossible to imagine without great faith in the power of chance. Biologist Richard Goldschmidt of the University of California claimed that it could not have happened in that way. He asserted that Darwinian evolution failed to explain the human ear and many other structures and organs of living creatures.[55]

8. Does the fossil evidence show gradual evolution of reptiles into birds?

Answer: The gap between the thecodont reptiles and the birds is said to be a period of about eighty million years with only the fossil *Archaeopteryx* to offer as an intermediate.[56] While this creature had some characters considered to be reptilian, it was fully feathered and definitely a bird. Some of the allegedly reptilian features of *Archaeopteryx* are possessed by some modern birds, and others are not found in certain modern reptiles.[57] Thousands of intermediate forms would surely have been required to originate feathers by evolution and transform reptiles into birds. But fossil evidence for the intermediate forms is missing.[58] Another problem is the fact that reptile lungs contain millions of tiny air sacs as do mammal lungs. But bird lungs have tubes rather than sacs.[59] How could a creature with a lung made half of sacs and half of tubes survive?

9. Does fossil evidence show the gradual evolution of the power of flight?

Answer: The evidence is nonexistent also for the origin of the other three types of flying animals. The flying insects were always flying insects,[60] the now extinct flying reptiles have no transitional fossils connecting them to non-flying reptiles,[61] and the flying mammals (bats) were always well-engineered bats.[62] Flying creatures were apparently designed to fly from the very beginning, just as reported in the Biblical record of creation.

10. Have not many gaps admitted by Darwin in the fossil record now been filled in with new finds of fossil "missing links?"

Answer: If anything, since Darwin's time the existence of the gaps in the fossil record has become more pronounced, accentuated by a century of largely fruitless search.[63]

11. Have not scientists traced the evolutionary tree connecting all forms of life, from single-celled forms to the present complex forms?

Answer: Examination of a standard evolutionary text such as *Vertebrate Paleontology* by Romer (ref. 46) reveals that the supposed evolutionary tree is actually a bundle of disconnected twigs. The charts in Romer's book are filled with dotted lines, both between the major groups and within these groups. The charts in *The Fossil Record* (ref. 52), give the same picture. The intermediate forms are missing, the twigs without connections to the branches and the branches disconnected from the roots.[64] This condition is characteristic in the case of both fossil and living kinds of plants and animals. And the plant fossil record is said to be even worse for evolutionary theory than the animal record.[65] Zoologist Bolton Davidheiser in his book, *Evolution and Christian Faith,* cites eighty statements in the technical literature in which evolutionary scientists admit they do not know the origin of eighty different kinds of animals and plants.[66]

12. What about the fossil horses? Aren't they proof of evolution?

Answer: The fossil horse series illustrated in school and college textbooks and in museums are highly simplified and rather misleading. They make the theory of horse evolution seem very neat, all cut-and-dried. Actually there are important problems with the theory and some disagreement even among evolutionary scientists.[67]

a. A complete series of horse fossils is not found in any one place in the world arranged in rock strata in proper evolutionary order from bottom to top. The fossils are found in widely separated places on the earth.

b. The currently accepted sequence of fossils starts in North America, then jumps to Europe and back to America again. But there is still differing opinion on whether one of the jumps was from America to Europe or vice versa. Many different evolutionary histories for horses have been proposed.

c. *Hyracotherium* (eohippus), supposedly the earliest, beginning member of the horse evolution series, is not connected by intermediate fossil forms to the condylarths from which it supposedly evolved.[68]

d. The first three supposed horse genera,

found in rocks classified as Eocene, are named *Hyracotherium, Orohippus,* and *Epihippus* and they are said to have evolved in that order. However, the average size of these creatures, sometimes called "old horses," decreases along the series, which is contradictory to the normal evolutionary rule, and they were all no larger than a fox.[69] These three genera could be considered to be members of an originally created biblical "kind."

e. Between *Epihippus* and *Mesohippus,* the next genus in the horse series, there is a considerable gap.[70] The size increases about 50 percent and the number of toes on the front feet decreases from four to three. The series of genera, *Mesohippus, Miohippus,* and *Parahippus,* sometimes called the (small) "new horses," were three-toed animals much more similar in appearance to modern horses than the previous group discussed. These, perhaps, were members of another created kind.

f. *Merychippus,* The next genus in the supposed horse evolution series, and the first of the (large) "new horses," was about 50 percent larger than the group of genera just discussed. It was three-toed, but the two side toes on each foot were quite small and unimportant, and the animal looked very horselike. *Pliohippus,* the next genus in the series, was a one-toed horse. These animals had some characteristics of skeleton and teeth which differed from modern horses, but they may, perhaps, be classified as members of the same original created kind.

g. In summary, the so-called fossil horse series actually appears to be three groups of genera. The first in the series has no connection by fossil intermediates to the supposed ancestors. The three groups may well have no connection one with the other, and the overall fossil horse data can be fitted into the framework of the biblical creation model. There is no need to assume that horses were evolved rather than created.

h. Two mysteries surround the theory of horse evolution. The first arises from the fact that the brain of little *Hyracotherium* is simple and smooth, whereas that of true horse, *Equus,* has on its outer surface a complex pattern of folds and fissures.[71] Cattle brains are quite similar and equally complex and have an almost identical pattern of fissures. Cattle and *Hyracotherium* supposedly evolved from a common ancestor which had a much simpler pattern of fissures. Therefore, it must be assumed that parallel evolution by accidental processes produced the same complex brain pattern in cattle and horses. Such a tale is difficult to swallow. Intelligent, purposeful creation provides a more rational explanation.

The second mystery lies in the massive extinction of horses in the Americas. What happened to them no one knows, and scientists can only speculate. Is the solution to this mystery to be found in the Genesis record of a universal flood which destroyed all air breathing land animals except those on the ark with believing Noah and his family?

i. We would conclude that the fossil evidence which is supposed to prove horse evolution can also be fitted into the framwork of Biblical creation. It all depends upon ones world view. The faith of atheistic materialism leads one to evolved horses. The faith of Biblical theism leads to created horses.

13. Do fossils of now extinct creatures such as dinosaurs show that evolution has occurred?

Answer: The fact that dinosaurs once lived and are now extinct is no proof of evolution.

Such fossils merely show us that certain species once living were destroyed and became extinct. Theorists have been able to reach no general agreement on the cause or causes of extinction. The theories on this subject are numerous and sometimes very imaginative.[72] Since most fossils are found in sedimentary rocks and show signs of catastrophic burial, (See chapter H) they seem to point to a global flood as the principal cause of extinction.

14. I thought that dinosaurs became extinct long before man appeared on the earth.

Answer: Recent explorations in Texas

revealed human footprints in the same limestone surface with dinosaur tracks. They must have lived on earth at the same time, just as the Bible implies.

If the flood-geology interpretation of geological strata is correct, all or most dinosaurs became extinct at the time of the Flood. Until that time, then, man and dinosaurs lived on the earth at the same time. Is there any evidence outside of the Bible to support this view? Yes, there is. It is well known that along the Paluxy River in Texas many dinosaur footprints have been found in limestone strata classified as Cretaceous. Not so well known is the fact that for about fifty years human footprints have been reported in the same strata.

In 1970 the Creation-Science Research Center, the Creation Research Society, and Films for Christ conducted a joint investigation of these reported finds. The course of this investigation and the results are recorded in the color-sound film, *Footprints in Stone.* Many footprints as well as series of prints were discovered and filmed. There were children's prints, those of normal adults, and also very large prints up to eighteen inches long. The human characteristics were unmistakable. Some human prints were found overlapping large three-toed dinosaur prints. One series of footprints was followed to the river bank, where it disappeared beneath an overlying stratum of limestone. This limestone layer was stripped off with a bulldozer. There, revealed to modern eyes for the first time, was the continuing series of footprints. They had to be real, bona fide, impossible to fake.

One evolutionary geologist from the University of Texas at El Paso agreed that the prints appeared to be human. But since the concept that man lived with dinosaurs is incompatible with traditional historical geology and the theory of evolution, he felt it could not be so. It is interesting that no secular university has conducted a serious investigation of these man-like footprints in the dinosaur country along the Paluxy River of Texas. The on-the-scene documentary film, *Footprints in Stone,* a most persuasive production, is available for rental or purchase.[73]

15. Does the fossil record show living forms continually changing, with ancient types becoming extinct and replaced by new, different forms?

Answer: This is the evolutionary view, but there are quite a few so-called "living fossils", plants and animals living today which are essentially or almost identical to fossils found in rocks supposedly millions of years old. If all populations have continually been evolving, why are these creatures so constant and unchanging? The following are examples of living fossils with their alleged ages in millions of years:[74] bat (50MY),[75] tuatara (135MY),[76] neopilina (500MY),[77] cockroach (250 MY),[78] dragonfly (170MY),[79] starfish (500MY),[80] metasequoia tree (60MY),[81] Ginkgo tree (200MY),[82] cycad tree (225MY),[83] coelacanth fish (65MY),[84] Port Jackson shark (180MY),[85] sea lilly (160MY),[86] sea urchin (100MY),[87] Vampyroteuthis or squid-octopus (200MY).[88] It does appear incredible, if evolution is a true theory, that these creatures could so long have remained relatively unchanged. Perhaps the millions of years and the theory of evolution are both myths created by materialists to get rid of the Creator.

Gould, Stephen J., *Natural History, 86,* June-July, 1977, pp. 22, 24. . . . The fossil record with its abrupt transitions offers no support for gradual change. . . . All paleontologists know that the fossil record contains precious little in the way of intermediate forms; transitions between major groups are characteristically abrupt.

Corner, E.J.H., in *Contemporary Botanical Thought*, MacLeod and Cobley, eds., (London: Oliver & Boyd, 1961), pp. 96, 97. . . . I still think that, to the unprejudiced, the fossil record of plants is in favour of special creation. . . . Textbooks hoodwink. A series of more and more complicated plants is introduced . . . and examples are added ecclectically in support of one or another theory—and that is held to be a presentation of evolution.

White, Harold, *Proc. of the Linnaean Soc. of London, 177,* Jan. 1966, p. 7. . . . I have often thought how little I should like to have to prove organic evolution in a court of law.

Section G

FOSSIL MAN?

1. Do fossils of prehistoric man-like creatures show the evolution of apes to ape-like cave men to modern man?

Answer: Most of the "fossil men" were merely animals having no connection to the human race. Some of the fossils which are true human remains completely contradict the theory of human evolution. Several famous fossil finds were frauds upon the scientific world and the public.

If human evolution from ape to primitive cave man to modern man had really occurred, the fossils should have been found in that order from the lower to the higher rock strata or layers. That is, the more ape-like fossils should be found in rocks dated as older, and the fossils more similar to modern man should be found in the rocks dated younger. Contrary to what is presented in the textbooks and newspapers and television, the actual picture is not so simple.

2. Have true human fossils been found in the wrong strata to support the evolution theory?

Answer: Yes, but for the most part fossil finds not fitting the theory are ignored. Fossil remains the same or essentially the same as modern man which were found buried very deep or in strata dated very old have been ignored and are no longer reported to the public. Examples are the Calaveras, Castenedolo, and Olmo skulls (See table). British anthropologist Sir Arthur Keith in his book, *The Antiquity of Man,* described these and other like fossil man finds in detail and stated that they would have been readily accepted by scientists if it were not for the fact that these fossils, because of their locations in the strata, contradicted the accepted theory of human evolution.

3. Are there cases of fraud in the history of fossil man finds?

Answer: The Piltdown fossil found in England in 1912 was shown in 1953 to be a cleverly contrived hoax.[89] The greater part of the scientific world had accepted the fraud for forty years.

4. Have the various fossil candidates for a place in our human ancestry stood the test of time?

Answer: One by one, the various fossil man finds have flashed across the front pages of the newspapers and been the subject of many scientific studies and reports in technical journals, only to be at last discredited and forgotten. replaced by newer finds which also come to sad endings. For example. . .

5. Were the Neanderthal people really crude, hunched over, bestial creatures that evolved into modern man?

Answer: For many decades most anthropologists were completely wrong. Neanderthal has now been found to have been an intelligent human being who walked perfectly upright, not a stupid, hunched-over half-ape-half-man.

Human remains were discovered in 1856 in a cave in the Neander Valley, which was the source of the name, Neanderthal.[90] A Neanderthal skeleton found in 1908 was the model for textbook drawings and museum displays of Neanderthal men and family groups used for many decades afterwards. These illustrations portrayed them with bestial features, bull necks, haunched-over posture, and knees which could not be straightened. In 1956 respected evolutionary scientists reexamined the bones and concluded that they were of an individual who suffered from severe skeletal malformation resulting from rickets and arthritis. They determined that Neanderthals walked as upright as we do and that, dressed in modern clothes, they would probably draw no special attention among the crowds in the New York subway.[91] Other evidence shows that the Neanderthal people were intelligent, skillful, artistic people who believed in life after death. They were true men, *Homo sapiens*.

We would speculate that the Neanderthals were a branch of Adam's race which through the effects of environment and other factors suffered some changes in the shape of their skulls, a character of the human body which is actually somewhat plastic. In fact, a number of the man fossils may represent peoples which had suffered degeneration as the result of sin, crude pioneer living conditions, and inbreeding in small frontier population groups after the Flood.

6. What became of Java Man?

Answer: Java Man, also called *Pithecanthropus erectus* or, more recently, *Homo erectus,* discovered in 1891, apparently represents a case either of fraud or misinterpretation, or both. The finder, Eugene Dubois, admitted some thirty years later that he had found in 1889 at Wadjak, Java, a true human skull of very large brain capacity. It was located in a layer not necessarily younger than that in which the *Pithecanthropus* bones were found. Some authorities always considered that the *Pithecanthropus* skull belonged to an animal, and in 1936 Dubois himself concluded that the creature was actually a giant gibbon.[92, 93]

7. What happeneded to Peking Man?

Answer: The Peking Man or *Sinanthropus* fossils reportedly found in 1928 and succeeding years were never permitted to leave China. Only plaster casts and models were exported. The bones disappeared during World War II under somewhat mysterious circumstances. A jumble of conflicting scientific reports were published over a period of a quarter of a century. When carefully compared, these reports show that Peking Man was an animal, probably a large monkey or baboon, not a man. Moreover, true human skulls were found in the same huge ash pit. Two outside authorities were permitted to examine the bones and the discovery site. Abbé Breuil described the massive lime burning pits in which the bones were found and also raised serious question about the theory that the *Sinanthropus* fossils were ancestral to humans. Later Marcellin Boule, international authority on fossil skulls, made a careful study of the bones and the site and published his conclusion that *Sinanthropus* was an animal which was eaten by the true men who had manufactured lime at the site. There is much appearance of fraud in the history of the Peking fossils.[94]

The Peking Man story closes mysteriously. Toward the close of the Second World War the bones supposedly were packed in Marine lockers to be transported to America for safety. They disappeared without a trace. Perhaps it was deemed prudent to make it impossible ever to compare the actual bones with the plaster models and casts and the written descriptions which had been distributed around the world.

8. Has *Australopithecus* also come to a tragic end?

Answer: Over the past two decades *Australopithecus* has been the major contender for leading place among human ancestors. But, alas, he too is fallen. Louis Leakey was the principal *Australopithecine* researcher at the time of his death in 1971. A year later his son, Richard, announced a find which was much more human and which he dated a million years older, thus throwing the accepted views of human evolution into confusion. Furthermore, the elder Leakey himself had found evidence of a human living area at a lower level than the *Australopithecus* remains, and son Richard had published in 1971 evidence that the creature was a "knucklewalker" not unlike the living African apes which are long-armed, short-legged knucklewalkers.[95]

The final blow came in 1975 with the publication of a quantitative computerized study and comparison of the bones of apes, *Australopithecus,* and man. The computer classified bones of the three kinds of creatures in three separate groups and showed that they represent different kinds of locomotion. The conclusion from all this is that *Australopithecus* has now gone completely down the drain as far as human ancestry is concerned.[96] Now who will step up to be the next candidate?

9. The table below, mainly adapted from Cousins,[92] illustrates the problem of human fossils found in strata dated older than the strata containing their alleged ancestors.

Specimen	Type or Form	Stratum	Assigned Age MY=million yrs.
Neanderthal (Spy skull)	Homo sapiens	Late Pleistocene	35-70,000 yrs.
Swanscombe	Homo sapiens	Middle Pleist.	0.25-0.6 MY
Pithecanthropus erectus	Ape-like	Middle Pleist.	0.5 MY
Sinanthropus	Ape-like	Middle Pleist.	0.5 MY
Australopithecus africanus	Ape-like	Early Pleist.	
Zinjanthropus	Ape-like	Early Pleist	1.75 MY
Skull 1470[97] (Leaky, 1972)	Man-like	Pliocene	2.8 MY
Ethiopean jaw[98] (Taieb & Johanson, 1974)	Man-like	Pliocene	3-4 MY
Castenedolo	Homo sapiens	Pliocene	2 MY+ ?
Calaveras	Homo sapiens	Pliocene	2 MY+ ?

10. What is the meaning of the recent finds by Richard Leaky (Skull 1470)[97] and by Taieb and Johanson,[98] bones of rather more man-like types in rock strata said to date at 2.8 to 4 million years old?

Answer: Both Leaky and the Taieb-Johanson team have claimed that their fossil finds make all previous theories of human evolution obsolete, but they have little to offer as substitute theories. Of course, like other evolutionary paleontologists, they have chosen to ignore the Castenedolo and Calaveras skulls as well as other human fossils which have been discovered during the past century and left, forgotten, on dusty museum shelves. Could it be that the Biblical data on the origin of man is not obsolete after all?

11. What have recent studies of human genetic variability revealed about the origin of the human race?

Answer: The limited number of different forms (called alleles) of particular genes established now among the human population suggests that the human race may have expanded from a very small population only thousands (not millions) of years ago.[99] The geographic variations of human genes indicate that the human race radiated from the Middle East.[99] Both of these conclusions are in agreement with the Biblical record of creation and a catastrophic flood thousands of years ago, followed by the radiation of Noah's descendants from the region of Turkey.

12. Have not archaeologists found that human culture evolved through successive stages from cave dwellers to nomadic hunters to farm village dwellers and, finally, to builders of great city-states?

Answer: Actually, no. The evidence from archaeology shows the sudden appearance of the advanced Sumerian civilization without signs of its slow evolution upward from cave men. The observed facts really fit what the Bible says.

Archaeologists have discovered little or no evidence of historical roots for the first great civilization in Sumeria.[100] When the Sumerian people appeared in the Mesopotamian River Valley, they brought with them metallurgy, art, and the potter's wheel, as well as writing, all in a highly developed state. Archaeologist C. Leonard Wooley estimated at least a thousand years of cultural development before this point in their history, but where this happened he does not know. Thus, evidence for the slow evolution of civilization is lacking, but the facts do fit the Biblical record. In Genesis 4 we are told of the early development of cities and of technology and art — metallurgy, domesticated animals, and musical instruments. Genesis 6-9 tells of a family of eight people who survived the judgment of the global flood and who must have preserved much knowledge of the former culture and technology. Thus as the race became reestablished in the post-flood world and population began again to swell, civilizations could flower rapidly without long evolutionary growth. Archaeology supports this Biblical model for the origin of ancient civilizations.

13. Isn't modern man much more intelligent than ancient man?

Answer: There is no evidence for the evolution of human intelligence. Since man is a cultural being, modern men have the benefit of the knowledge discovered by earlier generations, but not a higher level of intelligence.

This is the view of the noted French social anthropologist, Claude Levi-Strauss[101] The ancient peoples were highly intelligent, accomplishing marvelous feats of architectural design and engineering without the help of the refined instruments and massive power machinery available today. Lacking mass-produced books, and having no computerized data banks, they had to make much greater use of their memories than do moderns. If it were possible to interview the ancients, we might perceive that human intelligence has actually degenerated. On the other hand, modern man knows more than the ancient peoples and possesses more advanced technology because man is a cultural being. Each successive generation stands upon the shoulders, so to speak, of former generations, benefiting from their labors and scholarship. Man is the only cultural being, for he alone can bind time, preserving present advances for the future and drawing upon past accomplishments for the present.

14. Isn't the idea of Eve's being created from Adam's rib pretty weird?

Answer: The creation of Eve from Adam's side is not really so wild as some people think.

This teaching of the Bible agrees with the modern science of genetics.

Few people realize that the Genesis account of the creation of woman from man accords with modern knowledge of genetics which was unknown to Moses.[102] In humans, sex is determined by the two sex chromosomes. The female has in each body cell two X chromosomes, whereas the male has an X and a Y. Thus, if the female had been created first, it would not have been possible to create the first man from genetic material entirely related to the woman. This is because God in making Adam would have had to create Y chromosomes, for Eve had no Y chromosomes in her cells. As a consequence the resulting race would have been a hybrid race. But because man was created first, woman and man could be completely related to each other. This unity of the race in Adam is theologically very important, for we all sinned in Adam and fell with him in his first transgression. The Redeemer of the fallen race, Jesus Christ, receiving human nature by a miraculous conception in the womb of the Virgin Mary, became — and we say it reverently — a hybrid being, the God-Man. And all those who believe in Him are united with and in Him and receive a new nature, becoming the children of God by a spiritual rebirth. (See Romans 5-6, I Cor. 15)

15. But where did Adam and Eve's sons, Cain and Seth, get their wives?

Answer: They married their sisters in order to begin the expansion of the human race.

We are told in Genesis 5:4 that after the birth of Seth, Adam begat sons and daughters. So the answer to the question is that Cain married one of his sisters. But some will complain, that's not safe, is it? Is not close intermarriage dangerous because of possible genetic problems? Yes, it is now, when all humans carry in their chromosomes a "genetic load" of bad mutations. In the early years of the race few bad mutations had accumulated in the human gene pool, so close intermarriage was not genetically dangerous. In the beginning history of the race close intermarriage was necessary. But when the race had multiplied sufficiently, close intermarriage was forbidden.

16. Does human nature provide evidence for the existence of the Creator God of the Bible?

Answer: The essential attributes of human nature are intellect, affections (i.e., our feelings of love, fear, compassion, etc.), moral capacity, and will. The human body is composed of the material atoms found in the dust of the earth, just as the Bible teaches. But is there any scientific evidence that atoms or molecules have any of the four attributes of human nature listed above, or that chemical reactions can give these attributes to dust? No such evidence exists. There is no reason to believe that non-living matter thinks, has feelings, has any sense of moral responsibility, or exercises will, or that chemical reactions can make an organism that does. Personal nature, therefore, must have come from a higher personal spiritual Source, not from an impersonal material source. This conclusion from the scientific evidence is just what the Bible teaches. We were created in the image of the infinite-personal Spirit, God the Creator.

Leakey, L.S.B., in *Evolution After Darwin*, Vol. 2, Sol Tax, ed. (Univ. of Chicago Press, 1960), p. 23.

Almost all anatomists and paleoanthropologists are, however, in general agreement that the roots of the human stock must be sought in the group . . . Australopithecinae.

Oxnard, C. E., *Nature, 258,* 4 Dec. 1975, pp. 389, 394. . . . The genus *Homo* may, in fact, be so ancient as to parallel entirely the genus *Australopithecus* thus denying the latter a direct place in the human lineage. . . . If these estimates are true, then the possibility that any of the australopithecines is a part of human ancestry recedes. . . . We may well have to accept that it is rather unlikely that any of the australopithecines, including *"Homo habilis"* and *"Homo africanus"* can have had any direct phylogenetic link with the genus *Homo* except perhaps at earlier times.

Zuckerman, Sir Solly, *Beyond the Ivory Tower* (New York: Taplinger Pub. Co., 1971), p. 64. . . . The record is so astonishing that it is legitimate to ask whether much science is yet to be found in this field at all.

Section H

FOSSILS AND GEOLOGY — SLOW OR FAST?

1. How are fossils formed, according to the theory of uniformitarian historical geology?

Answer: The accepted view is that fossils were the result of processes pretty much like those taking place on earth today.

The term "uniformitarian" refers to the idea that geological processes such as erosion, sedimentation, and earth movements have remained pretty much the same in character and rate for most of earth history. Fossils are generally found encased in sedimentary rock, that is, rock which was deposited by water in the form of loose sediments which were then compressed and cemented to form solid stone. Since in the traditional theory most sediments were laid down very slowly, those plants and animals which left fossils must for the most part have been covered and fossilized slowly. Also, the vast deposits of the fossil fuel, coal, supposedly were formed from forests and peat bogs which grew slowly, died, were covered with sediments slowly in the same location in which they grew, and were finally compressed to make coal.

2. Where can fossils be seen being formed today?

Answer: Present earth conditions are not producing fossils such as are found in abundance in fossil bearing rocks.

Fossils are not today observed being formed anywhere on the earth by the gradual processes just described. When plants and animals die they are immediately attacked by scavengers, fungi, and bacteria, which destroy them before they can be buried by sediments and fossilized. Any appreciable formation of fossils apparently requires sudden entrapment and rapid burial, a catastrophic process.[103]

3. Do fossils generally give the appearance of having been formed by slow or by rapid processes?

Answer: A great many facts point to rapid, catastrophic burial of the plants and animals which are found as fossils today.

a. Fossil caves, fissures, mass burial sites, and sedimentary strata discovered in Europe and America were jammed with masses of mixed bones of many sorts of animals from widely separated and differing climatic zones, for example: (1) Cumberland Cavern in Maryland, containing remains of animals from cold northern regions, warm, damp semi-tropical regions, and from more arid environments,[104] (2) Norfolk forest-beds in England, which contain remains of temperate zone plants, and large numbers of both northern cold-climate and tropical warm-climate animals, all mixed together,[104] (3) rock fissures in England and France contain masses of broken bones of many kinds of animals from both cold and temperate zones;[104] (4) the Baltic amber deposits and the Geiseltal lignite seams in Germany contain fossil insect and plant and animal remains which must have been collected by some cataclysmic process from different areas all over the earth, from near arctic to tropical zones, and transported from Africa, the East Indies, and South America to be dumped in northern Europe.[105]

b. Numerous fossil graveyards contain stupendous quantities of fossilized bones of many different kinds of animals thrown together in jumbled masses so to be explainable only in terms of catastrophic water action of vast proportions. These include the Agate Spring Quarry in Nebraska, the Rancho La Brea Pit in Southern California, the Siwalik Hills fossil beds in India, and the fossil fish graveyard strata of Lompoc, California, the Old Red Sandstone in Scotland,

and many other fish graveyards in Italy, Germany, Switzerland, etc.[106]

c. In many locations in the world are found extensive rock strata containing sometimes billions of fossilized animals, frequently densely packed together. These seem to suggest anything but slow, calm conditions of formation.

d. The vast coal deposits of the world generally do not give evidence of having been fossilized in the same location as the plants originally grew. Instead they appear to have been dumped into place by flood action.[107] Usually there is no evidence of a soil layer in which the supposed forest once grew. There are also many examples of "polystrate fossils," fossilized tree trunks which extend through a number of layers of sedimentary rock and coal, some of them up to one hundred or more feet in length.[108] These surely must have been covered up very rapidly in order to fossilize into coal before the termites got to them. Sometimes up to one hundred layers of coal separated by rock layers have been found at one location. These data strongly indicate that the vegetation was rafted in from other locations by great water currents, dumped, covered with sediments, and rapidly converted to coal by the pressure of deep overlying sediments. Pressure and heat can convert wood to coal in days,[109] and garbage can be changed to oil in less than an hour by a recently developed commercial process. The entire fossil record is much more easily understood to be the result of global flood action of great rapidity and violence, rather than of the relatively slow processes mostly observed today.

4. Are sixty-five petrified forests stacked one on top of the other at Specimen Ridge in Yellwstone Park? Does this famous geological formation represent a vast span of time?

Answer: Much evidence shows that the Specimen Ridge "fossil forests" are not the remains of forests which grew one on top of the other during long periods of time. Rather, it appears that trees from distant forests were ripped up and transported by water to be dumped at Specimen Ridge. The facts strongly indicate that the standard view long held by geologists is completely wrong.

The traditional view held by geologists is that the Yellowstone petrified tree formations represent many forests which grew one after the other. Each took hundreds of years to grow before it was buried by volcanic ash and slides of volcanic breccia (sharp-edged chunks of volcanic rock cemented to form a solid rock). Then another forest grew on top of it, only to suffer a similar fate, until perhaps as many as fifty or more forests had been buried and petrified. This explanation has been accepted without question for almost a century. However, recent detailed research has brought to light much evidence which contradicts the traditional view.

Dr. Harold Coffin has conducted careful

studies over a number of years on all aspects of the Yellowstone petrified tree formations. Some of the facts about these formations which do not fit the picture of forests' being buried where they grew are as follows:[110]

a. Tree roots abruptly terminating or broken.

b. Almost all trees completely stripped of bark or limbs.

c. Small trees upright, unbroken (a breccia flow would push them over).

d. Ring patterns of trees do not cross-match.

e. Both upright and prone trees lined up as if by water current.

f. No valid evidence of soil layers where trees grew.

g. Absolutely no evidence of animals found where soil layers should be; also, very few cones found.

h. Many examples of trees overlapping, with roots on one located at a level part-way up the trunk of another.

i. Broad leaves found where tree trunks are only conifers.

j. Pollen scarce and not of same kind as the tree trunks.

These and other facts strongly suggest that geologists have been wrong for a hundred years. The evidence better fits the view that trees were ripped up and transported by water from another location and dumped in place at the same time that repeated volcanic eruptions were layering the area with ash and breccia. The evidence supports the view that this happened rapidly, not slowly over periods of tens or hundreds of thousands of years.

5. Do the erosion and deposition of sediments seen today explain how sedimentary rock layers could have been formed in the past?

Answer: The major features of the sedimentary rock strata cannot be explained in terms of the processes seen on the earth today.[111]

a. Vast horizontal strata point to global flood.

In our southwestern states, where they are particularly well exposed by erosion, but also across the continent and everywhere in the world, thousands to hundreds of thousands of square miles of flat, horizontal strata, from a few feet to hundreds of feet thick are found. At no location on the earth may the production of similar sedimentary deposits of like extent be observed today. These formations are composed of sandstone, graywacke, shale, conglomerate, limestone and other types of rock, some of them blanketing whole continents and extending for thousands of miles. For example, the St. Peter sandstone has been traced in twenty states from California to Vermont.[112] The Shinarump conglomerate in the Southwest covers some 125,000 square miles,[113] and another conglomerate blanket is reported to extend from New Mexico to Saskatchewan and Alberta.[114]

A continental blanket of sandstone required a steadily flowing current traversing thousands of miles to separate the sand from silt and gravel before deposition. A continental blanket of conglomerate required a continent-sized maelstrom of water in violent, chaotic motion to dump an ungraded mixture of materials of all sizes across thousands of miles of terrain. Tremendous water action such as that which would be produced by a global flood seems to offer the only reasonable explanation for the observed facts.

b. Finer structure of strata explained in terms of the flood.

The finer structure of sedimentary strata, called stratification, is also difficult to explain satisfactorily in terms of the more or less gradual processes observed in action today. The four common types of stratification seem to be more easily explainable in terms of the kind of very rapid water action which a global flood would have produced. Simple lamination,[115] cross lamination and cross bedding,[115] ripple lamination,[116] and graded bedding[117] all are equally well or better explained in terms of the global flood hypothesis than they are in terms of traditional uniformitarian concepts of historical geology.

c. Composition of sedimentary strata unexplained.

Not only the structure, but also the composition of these rock formations bears witness to catastrophic deposition of most of the sedimentary rocks in the earth's crust. Limestone,[118] dolostone (limestone containing much magnesium carbonate,[119]) cherts (flint-like stone),[120] graywacke,[121] and "evaporites" (such as gypsum or rock salt)[122] cannot be suitably explained in terms of processes observed on the earth today. In fact, they apparently were formed by water currents and oceanic chemical reactions on a scale which is incomprehensible in terms of present earth activities. Geologists cannot agree on the explanations for these types, and there is much mystery still, but one thing is certain: a global flood seems to offer the best possibility ultimately of explaining all of the observed facts.

d. Vast volcanic lava outpourings.

In the states of the Pacific Northwest, in India and elsewhere, hundreds of thousands of square miles of territory were engulfed by floods of basaltic lava which must have flowed like rivers and stacked up thousands of feet thick. Volcanic action on this scale is unheard of in the modern world.[123]

e. Earth's crust violently altered.

Many other characteristics of the rock structures of the earth's crust suggest catastrophic activity on a giant scale not seen today. Large scale folding, faulting, and uplifting and sinking are examples. Great river canyons in the ocean bottoms and shallow water deposits on the sea floor indicate that the oceans may have been thousands of feet lower than at present.[124] Striking evidence from archaeology indicates that the Andes and Himalaya mountain chains were pushed up thousands of feet in historic times.[125]

f. Mt. Ararat submerged.

On Mount Ararat, a volcanic mountain complex, pillow lava is found at the 14,000 foot level.[126] Pillow lava has been extruded under water and is recognized by its high glass content caused by very rapid cooling. All of the lavas examined on Ararat by geologist Clifford Burdick were extruded under water. Ararat apparently was submerged in water to above the present 14,000 foot level. Thus the entire world must have been inundated at the same time.

All of these evidences from geology and many more point to a violent, catastrophic past history of the earth. They imply a global flood with associated volcanic and mountain building activity which changed the face of the earth.

6. Are the rock layers and their embedded fossils always found in the same order, with simple fossils on the bottom and complex ones on the top, in the same order in which evolution is said to have occurred?

Answer: There are many places on the earth where rock strata and fossils are found in the reverse order from that predicted by evolutionary theory. Many of these are very difficult for geologists to explain. If they cannot be explained, the history of the evolutionary change gets turned around backwards, which is embarrassing for the theory.

In our files are references from the scientific literature to hundreds of such reversals. Often a reversal can be shown to be caused by the overturning of a fold in intensely deformed sediments. But in other cases folding cannot explain it. The stock explanation in such cases is that a "thrust fault" allowed older strata containing simple fossils to be slid out on top of younger rocks containing complex fossils.

The most famous example of a so-called thrust fault is the Lewis Overthrust, covering some 13,000 square miles of mountain ranges in Montana and British Columbia. However, the physical evidence that an entire system of mountain ranges slid thirty to sixty miles out on top of underlying strata is absent. A thick layer of ground-up mixed rock from the two layers should be found between the upper and lower sections of the supposed overthrust structure. It is nowhere to be found. The rock layers appear to have been deposited one on top of the other in the normal manner. If the simple algae fossils in the upper layers (rocks classified as Precambrian dolomite) were not supposedly 400 million years older than the complex marine fossils in the lower layers (classified as Cretaceous shale), the idea of a thrust fault in that area would never have occurred to geologists. But the theory of evolution must be saved at any cost, so heroic feats of geological imagination are performed to explain away the obvious facts.[127]

Another related type of evidence which is embarrassing to the defenders of the traditional geological claim is the discovery of many different kinds of pollen in Precambrian rocks and also the numerous findings of fossils of a large assortment of species of woody stemmed plants in rocks classified as early Paleozoic.[128] These data would have flowering and woody plants appearing on the earth up to a half billion years too early to fit in the theory of evolution.

The important point is that fossils appear in the wrong order in many places, but the order of the fossils in the rock layers is one of the principal evidences offered that evolution from simple creatures to complex actually took place. In addition, rock strata corresponding to hundreds of millions of years of theoretical time are often missing. Perhaps the evolution and the years are really imagination.

7. What is the Biblical picture of earth history?[129]

Answer: While the Bible does not give a detailed picture of earth history, it does provide a framework for interpreting the fossils and rock strata. The opening verses of the Bible indicate that in its original form the earth was surrounded by water, perhaps in violent activity. Thus the earliest sedimentary rocks formed would be devoid of fossils, for life had not yet been created.

This was on the first day of creation, ten thousand or so years ago. On the third day of creation God lifted the original continental mass from beneath the waters. Probably until the Flood, thousands of years later, the level of the dry land was generally lower and more even than today, the mountains much lower than today's mountain ranges. Sea level was considerably lower and the land surface therefore greater than at present. When God judged the sinful human race with the Flood of Noah, rain fell and perhaps additional water was brought to earth by special divine provision, and juvenile waters poured out through fissures in the crust of the earth in tremendous volume. In addition, the ocean bottoms and the land surface may have changed their relative levels, the former rising and latter sinking.

The effect was to cover the land surface totally and subject the entire earth to the action of global water currents and wave action of cataclysmic violence. The land surface was deeply eroded and every living thing swept away. The sediments with the dead plants and animals were deposited to form the strata observed today with their content of fossils. Great earth upheavals, volcanic activity, vast lava outpourings, and rapid mountain building accompanied the latter part of the Flood and continued for centuries afterwards on a diminishing scale. Perhaps in the centuries after the Flood the original land mass began to separate into the present continents.

Genesis 10:25 may refer to a crucial stage in this separation, as well as to the division of the nations at Babel. In the post-flood period glaciers advanced and retreated. Gradually the conditions on the earth stabilized, the land and seas were filled again with living creatures, and the descendants of Noah spread around the world.

Gregory, Herbert E., *U.S. Geological Survey Prof. Paper 188,* 1938, p. 49. . . . The physiographic conditions under which the Shinarump (conglomerate blanket, ed.) was deposited are difficult to visualize. What conditions could be so persistent and so uniform as to permit the deposition of a thin sheet of material essentially alike over thousands of square miles in Utah, Arizona, and Nevada?

Dunbar, C. O., and Rogers, John, *Principles of Stratigraphy* (New York: John Wiley & Sons, 1957), p. 245. . . . Chert has been the subject of as much controversy as dolostone, and for about the same reasons: neither is known to be forming anywhere today, and the chemistry of formation of each presents difficult problems.

Nilsson, Heribert, *Synthetische Artbildung* (Lund, Sweden: Gleerup, 1954; reprint by Evolution Protest Movement of N. America, Victoria, B.C., 1973), pp. 1194-1195.

Let us study . . . the formation of amber. The largest deposits . . . in East Prussia . . . estimated . . . at 5 milliard (5×10^9) kilos. . . . In the pieces of amber . . . insects are of modern types. . . . It is then quite astounding to find that they belong to all regions of the earth, not only to the Palaearctic region. . . . Typically tropical species occur, from the Old World as well as from the New. The same is the case with the plant fragments. Leaves of tropical trees from East India, Borneo, Australia and South America are mixed with those from . . . homely shrubs. . . . The genus *Pinus* . . . needles . . . are Japanese or North American. . . . The geological and palaeobiological facts concerning the layers of amber are impossible to understand unless the explanation is accepted that they are the final result of an allochtonous process, including the whole earth. . . . Exactly the same picture as the one just given is offered by the well-known studies of certain fossil-carrying strata of the lignite in Geiseltal (Germany).

Section I

EVIDENCES FOR EVOLUTION — CAN CREATION EXPLAIN THEM?

1. Do the many similarities among monkeys, apes, and humans lend strong support to the theory that they are connected by evolutionary descent?

Answer: Such similarities do not prove, for example, that apes, monkeys, and men are descended from a common ancestor. The similarities can also be understood in terms of creation by a common Designer who used a basic plan with modifications.[130]

As an example, the basic plan for vertebrates is the quadrupedal (four-footed) design, because it is a practical arrangement in most instances. Consider, as an illustration, why all automobiles with few exceptions have four wheels. The reason is not that they all came from the same production line, but that four wheels is a good basic design. It can be modified to serve in a sports car (special suspension), jeep (four-wheel drive), passenger sedan (softer suspension), heavy duty truck (dual wheels), etc. The designers used modifications suited to specific purposes.

The bones of the different vertebrate animals generally correspond bone-for-bone, though they may have different functions in the different species. This is called homology, and it is considered to be evidence of descent from a common ancestor. Moreover, close similarity is interpreted as proving close evolutionary relationship. But as we said above, the creationist interpretation is equally valid.[130] Thus the forearms of the vertebrates have many forms, including the legs of salamanders, lizards, horses, apes, and men, as well as the wings of birds and bats. Each type of leg suits its purpose admirably, as would be expected of the work of an infinitely wise Creator. Furthermore, the actual evolutionary transition from a front leg designed for walking to the bat's wing, for example, has yet to be supported by fossil evidence. All fossil bats were well-engineered flying bats, and no poorly designed or intermediate ancestors have been found.[62] And in any event, it seems logically impossible to imagine how such a step-by-step transition could ever be accomplished so that the intermediate "bats in the making" could survive.

2. Is evolution supported by the fact that plants and animals can be classified into

groups, i.e., kingdoms, phyla, classes, orders, families, genera, and species?[131]

Answer: Some zoologists claim that they have a "natural" system of classification; that is, the system of classification supposedly parallels the evolutionary history, just the way they say it actually happened. They think, in other words, that in their system they see the family tree of all life. But we have already indicated that the tree is actually a bundle of disconnected twigs. Remember that Dr. Bolton Davidheiser in his book, *Evolution and Christian Faith,* was able to compile eighty quotations from the scientific literature in which scientists admitted that the ancestors or origins of eighty different groups of animals and plants are unknown.[66]

Furthermore, there has been much dispute over the proper placement of many animals and plants, and there has on occasion been a major shift in the placement of a group of organisms. For instance, the living ferns, once classified with the non-flowering plants, are now placed with the flowering plants.[132] Furthermore, there is neither a universally accepted definition for species nor agreement concerning how a species originates.[133]

In view of the many uncertainties connected with the classification system, it hardly can be relied upon as a solid support for any theory of evolution. The Christian view is that taxonomy (science of classification) systematizes and thereby promotes the knowledge and understanding of the Creator's handiwork.

3. Are there vestigial organs in some creatures which suggest that an evolutionary change from use to disuse has occurred?

Answer: Advancing knowledge of physiology has shown that most of the supposed vestigial organs are useful and even essential. If there are any true vestigial organs, they show the loss of structure and design, not the production of something new. But to support the theory of evolution, evidence for the production of new organs is required.

At one time one hundred and eighty vestigial organs or structures were listed for the human body. As the knowledge of physiology increased the list dwindled, uses being discovered for them, until now only a very few are offered as evidence for evolution.[134] One that is still sometimes suggested in biology texts is the human appendix. However, it is now thought that this organ, containing much lymphoid tissue, provides protection against infection, especially in infants.

Under certain conditions loss of organs or their function may possibly occur. Examples are insects, amphibians, or fish isolated in dark caves which have lost the power of sight, and certain insects on windy islands which have lost their wings because insects with large wings are easily blown off the island. However, the loss of a function or structure means the loss of genetic information from the gene pool, which is the reverse of what is required for evolution to occur.

4. Does the sequence of stages or forms of the human embryo display a history of evolution from a single cell to a worm-like creature to vertebrate fish to man?

Answer: This theory, called embryonic recapitulation (embryo retelling a story), is now almost entirely discarded by scientists because there are simply too many exceptions.[135]

Examples of exceptions include the following: in man the tongue develops before the teeth, vertebrate embryos form the heart before the rest of the circulatory system, some creatures are very similar in the adult stage but quite different in the egg or larval stages, there could be no ancestor corresponding to the formless jelly stage in the pupae of moths and butterflies, and the respiratory surface of the lung is the last to appear in the embryo, whereas it must have been present throughout the alleged history of evolution since the appearance of the land animals.

The so-called gill slits (actually pouches) and gill arches in the human embryo never have anything to do with respiration of the embryo, as they should if the recapitulation theory were valid. In the course of embryonic development they are incorporated into such organs as the Eustachian tube, the tympanic cavity of the middle ear, the palatine tonsils, the thymus, parathyroids, the carotid arteries, the subclavian artery, the aortic arch, and the ductus arteriosus. The theory that the gill pouches and gill arches are related to the gills of a fish ancestor is now discredited. It was based upon inaccurate and incomplete knowledge of the facts and upon inadequate understanding of the process by which the embryo develops.[136]

The logical understanding of the course of embryonic development is that a rational building plan is followed. The most complex structures generally start to appear first because they require more time for completion, and they must also be integrated into the other structures which develop later. Thus the facts agree quite logically with the creation viewpoint rather than with an evolutionary explanation.

5. Have studies of proteins in different kinds of plants and animals shown that proteins have evolved?

Answer: The information from protein studies can be understood in terms either of evolution or of creation. A series of proteins with systematic differences and similarities no more proves evolution than does a series of fossil skeletons or of living creatures. (See sections I-1, 2)

The protein subjected to the most thorough study of this kind is the protein chain of the enzyme, cytochrome c. This enzyme is important in the production of energy in all forms of life. It has been extracted from the cells of dozens of different species of plants and animals. The fundamental protein chain contains 104 amino acids, and the amino acid sequence (the order of the different amino acids) has been determined for each of the organisms. Although the enzyme does the same job in each organism, many differences have been found in the amino acid sequence. These differences have been analyzed in terms of the assumption that cytochrome c has evolved by mutations for a billion or so years, the rate of mutations being constant. If this were the case, then the number of differences in the cytochrome c molecules from two different species should be proportional to the length of time back to the separation of their two lines of ancestors from a common ancestor. These times are obtained from the theory of evolution and the time table of historical geology. Graphs of this relationship have been published which show a fair straightline relationship reflecting the number of differences proportional to the time.[137] This may be taken as circumstantial evidence for the theory that cytochrome c has evolved.

On the other hand, the data can be interpreted in accord with the Biblical creation model of origins.[138] The essential structure of the enzyme is the same for all of the species studied. For example, thirty-five of the 104 positions in the chain are invariant, being filled by the same amino acids in all of the species. Also, eleven of these unchangeable positions are together at the vital action center of the molecule where its enzyme function is performed. Apparently all of the cytochrome c molecules from the different species fold up into the same basic shape so that they can do the same job in cell chemistry. This supports the view that the design was created, not evolved. The differences may be understood as either the result of mutations, or created differences, or a combination of these. The fewer differences generally observed between similar creatures is in agreement with the creation model. The differences may in some cases be necessary or advantageous in a particular species. Finally, it should be pointed out that the theory and the graphs mentioned earlier have been sharply criticised by prominent evolutionary scientists. They charge that the time periods used in making the graphs are not really in line with current geological theory.[139] These critics seemingly are implying that the time periods assumed have been chosen so as to make the graph a straight line.

6. Wasn't Biblical creation defeated by evolutionary science in the famous Scopes Monkey Trial of 1925 in Dayton, Tennessee?

Answer: This infamous trial has been grossly misrepresented in the mass media for a half century with the result that very few people know the truth. For example, are you aware of the following facts?[140]

a. The idea of such a trial was originated in New York City by officers of the American Civil

Liberties Union. The legal defense was arranged and paid for by the ACLU and by members of the American Association for the Advancement of Science.

b. The ACLU released to the Tennessee newspapers a call for a teacher who would break the state law against teaching evolution.

c. The basic plans for John Scopes, a football coach and a substitute science teacher, to be the defendant, and for instigating the charges against him were made in an informal meeting in a Dayton drugstore.

d. John Scopes never testified in court to having violated the anti-evolution law by actually teaching evolution. He has since on at least four occasions apparently admitted that to the best of his knowledge he never did so. In other words, the ACLU, long noted for its defense of left-wing causes, perpetrated a fraud on the court and on the public.

e. Charles Darrow, agnostic lawyer for the defense, not only displayed ignorance concerning both the theory of evolution and the teachings of the Bible but also knowingly offered in evidence to the court bald-faced lies about the Bible. Moreover, throughout the trial he leveled a merciless barrage of insult and vilification against defense counsel William Jennings Bryan, who never responded in kind. How strange that the mass media have glorified and praised Darrow and ridiculed Bryan ever since.

f. William Jennings Bryan was actually the

hero of the trial, evidencing good understandings of the theory of evolution and its implications, of the teachings of the Bible, and of the relationship of the two. Yet for fifty years the mass media have portrayed him as a bigoted ignoramus.

g. There is evidence to support the contention that the ACLU leaders, in consort with various detractors of the Biblical Christian faith, arranged the Scopes Trial with the objective, as reported by one historian of the event, "to educate the public on evolution."

Sir Gavin de Beer, *Embryos and Ancestors*, revised ed. (Oxford Press, 1940-1954) pp. 6, 10. . . . Until recently the theory of recapitulation still had its ardent supporters. . . . It is characteristic of a slogan [Ontogeny recapitulates phylogeny] that it tends to be accepted uncritically and to die hard. . . . the prestige so long enjoyed by the theory of recapitulation had a great and, while it lasted, regrettable influence on the progress of embryology.

Paul Weatherwax, *Plant Biology* (Philadelphia: W. B. Saunders Co., 1942), p. 240. . . . Botanists still disagree widely on the proper grouping of many plants, but this is because they do not agree in their theories as to the origin of the differences which separate the groups.

Alfred Romer, *The Vertebrate Body* (Philadelphia: W. B. Saunders Co., 1962), p. 358. . . . This (the appendix, editor) is frequently cited as a vestigial organ supposedly proving something or other about evolution. This is not the case; a terminal appendix is a fairly common feature in the cecum of mammals, and is present in a host of primates and a number of rodents. Its major importance would appear to be in the financial support of the surgical profession.

Sir Gavin de Beer, *Homology, An Unsolved Problem,* Oxford Biology Readers, J. J. Head and O. E. Lowenstein, editors (Oxford Univ. Press, 1971), pp. 15. . . . *characters controlled by identical genes are not necessarily homologous. . . .*

The converse is no less instructive. . . . *homologous structures need not be controlled by identical genes,* and *homology of phenotypes does not imply similarity of genotypes.*

Section J

HOW OLD IS THE EARTH?[141] [142]

1. Haven't scientists proved that the earth is billions of years old?

Answer: As was shown in the first chapter where science was defined, the study of origins and earth prehistory is really beyond the powers of the scientific method. No humans were present to observe the events, and the events which occurred cannot be repeated experimentally. All of the evidence from the rocks is circumstantial and can be interpreted in various ways. Thus it is not possible to "prove" that the earth is billions of years old.

2. What are the requirements for a clock which measures time correctly?

Answer:

a. The clock must run at a known constant

rate. Nothing must happen to speed it up or slow it down.

b. The clock must be set correctly at the beginning of the time period being measured.

c. The clock must not be disturbed by movement of the hands during the time period being measured.

3. How do the radiometric dating methods for dating rocks work?

Answer: The radiometric dating methods rely upon radioactive elements contained in the rocks. An example is the uranium-238/lead-206 system. If a rock contains "parent" uranium-238 atoms, these continually decompose through a series of radioactive decompositions to produce, finally, "daughter" lead-206 atoms. It takes about 4.5 billion years for half of any quantity of U-238 atoms to decompose. This is called the "half-life" of U-238. If a sample of a rock is analyzed for its content of U-238 and Pb-206 atoms (Pb is the chemical symbol for lead), the ratio of lead to uranium atoms can be interpreted as a clock which tells how long ago the rock crystallized. The assumptions which must be made are: (a) the rock contained no daughter lead at time zero. (b) no parent uranium or daughter lead was either added to or taken from the rock since and (c) the rate of radioactive decomposition has not varied. Assuming these assumptions to be correct, if a sample of rock were found to contain the daughter and parent atoms in the ratio Pb-206/U-238=1/1, half of the uranium has decomposed to lead, so the rock is judged to have an age equal to one half-life or 4.5 billion years. If the ratio Pb-206/U-238=3/1, three quarters of the uranium has decomposed, so the rock is judged to have an age equal to two half-lives or 9 billion years.

4. Do the radiometric dating methods possess the three qualifications to measure time correctly?

Answer: The radiometric dating methods do not fulfill all of the requirements for a reliable clock.

a. The evidence generally supports the view that the rates of radioactive decay are constant within narrow limits. However, recent research suggests that special conditions may, perhaps, appreciably alter these rates.[143] The evidence for this is still not conclusive. It is possible that unusual conditions such as exposure to neutrino, neutron, or cosmic radiation could greatly have changed isotopic ratios or the rates at some time in the past.[144]

b. The daughter products of the various systems are all found widely distributed in the earth's crust, e.g., Pb-206, Pb-207, Pb-208, argon-40, and strontium-87. It is not possible to be sure that some daughter product atoms were not present in the rock at time zero.

c. Finally, all of the parent and daughter atoms can move through the rocks. Heating and deforming of rocks can cause these atoms to migrate, and water percolating through the rocks can transport these substances and redeposit them. These processes correspond to changing the setting of the clock hands. Not infrequently such resetting of the radiometric clocks is assumed in order to explain disagreements between different measurements of rock ages. The assumed resettings are referred to as "second" or "third events."[145]

From the above facts it can be seen that the radiometric dating methods do not in general fulfill all of the requirements for a reliable clock.

5. If the earth is really young, only thousands of years old, why do the radiometric methods usually give such large ages, millions or billions of years?

Answer: The half-lives of the parent atoms used in dating the rocks are very long, from hundreds of millions to billions of years. Since the daughter product atoms are found everywhere in the rocks — and they are equated to time — it should not be surprising to find that these methods yield large values for the age of the earth.

6. Are there special difficulties with some of the radiometric methods?[146]

Answer: Yes, all of the radiometric methods involve difficulties because of assumptions which are not necessarily true.

a. In the lead-uranium systems both uranium and lead can migrate easily in some rocks, and lead volatilizes and escapes as a vapor at relatively low temperatures.[147] Free neutrons can transform Pb-206 first to Pb-207 and then to Pb-208, thus tending to reset the clocks and throw thorium and uranium-lead clocks completely off, even to the point of wiping out geological time.[148] There is still a disagreement of 15 percent between the two preferred values for the U-238 decay constant.[149]

b. In the potassium/argon system argon is a gas which can escape from or migrate through the rocks. Potassium volatilizes easily, is easily leached by water,[150] and can migrate through the rocks under certain conditions. Furthermore, the value of the decay constant is still disputed, although the scientific community seems to be approaching agreement. Historically, the decay constants used for the various radiometric dating systems have been adjusted to obtain agreement between the results obtained.[151] In the potassium/argon system another adjustable "constant" called the branching ratio is also not accurately known and is adjusted to give acceptable results.[152]

Argon-40, the daughter substance, makes up about one percent of the atmosphere, which is therefore a possible source of contamination. This is corrected for by comparing the ratio argon-40/argon-36 in the rock with that in the atmosphere. However, since it is possible for argon-36 to be formed in the rocks by cosmic radiation, the correction may also be in error. Argon from the environment may be trapped in the rock by pressure and rapid cooling to give very high erroneous age results.[153] In view of these and other problems it is hardly surprising that the potassium/argon method can yield highly variable results, even among different minerals in the same rock.[154]

c. In the strontium/rubidium system the strontium-87 daughter atoms are very plentiful in the earth's crust. Rubidium-87 parent atoms can be leached out of the rock by water or volatilized by heat.

All of these special problems as well as others can produce erroneous results for the various radiometric dating systems.

7. Do the radiometric dating methods give consistent results?

Answer: Often they do not. Consider a few examples.[155]

a. Volcanic rocks on Reunion Island in the Indian Ocean yielded Pb-206/U-238 and Pb/206/Pb-207 ages from 2.2 to 4.5 billion years, but potassium/argon ages of 100,000 to 2 million years.[156]

b. Lunar soil from Apollo 11 gave ages by four different lead methods varying from 4.67 to 8.2 billion years and nearby rocks gave potassium/argon ages of around 2.3 billion years.[157] Certain Apollo 12 rocks gave strontium/rubidium and lead ages ranging from 2.3 to 4.9 billion years.[158] A certain rock from Apollo 16 gave lead ages from 7 to 18 billion years but was chemically treated until it yielded a corrected "acceptable" age of 3.8 billion years.[159]

c. Granite from the Black Hills gave strontium/rubidium and various lead system dates varying from 1.16 to 2.55 billion years.[160]

d. Certain Russian volcanic rocks gave ages from 50 million to 14.6 billion years, although they are believed to be only thousands of years old.[161]

e. Volcanic rocks from Hawaii extruded under water only 170 years ago gave potassium/argon ages from 160 million to almost 3 billion years.[153]

8. Do the radiometric dating systems offer a serious challenge to the Biblical chronology?

Answer: There is no question that a large body of radiometric age data has been organized to give strong apparent support to the view that the earth is billions of years old. On the other hand, discordances and anomalies such as those just listed suggest that another interpretation of the data is possible. In applying the radiometric methods to earth rocks, the time schedule currently accepted for the theory of evolutionary history controls which results are accepted and which are adjusted or discarded.[162] This in itself indicates that the radiometric dating systems are far from absolute.

While creationists have much scientific evidence in support of a young earth, (See Section K) it must be admitted that they do not have all the answers to the long-term radiometric dating systems. One important unsettled question is found in the isochrons devised for use with the strontium/rubidium system. An isochron is a graph used to interpret the measured amounts of strontium-87, rubidium-87, and strontium-86 in different crystals of a rock or in different rocks in a rock structure. If the graph is a good straight line, certain assumptions make it possible to calculate the original composition of the rock and its age. Sometimes such isochrons cannot be obtained, and sometimes they are very poor. On the other hand, occasionally very excellent isochrons are obtained which support a great age for the rock.[163] This, in our opinion, offers an important challenge for new research and analysis by creation scientists.

9. How ancient is life on earth according to the carbon-14 dating method?

Answer: A survey of the 15,000 radiocarbon dates published through the year 1969 in the publication, *Radiocarbon,* revealed the following significant facts:[164]

a. Of the dates of 9671 specimens of trees, animals, and man, only 1146 or about 12 percent have radiocarbon ages greater than 12,530 years.

b. Only three of the 15,000 reported ages are listed as "infinite."

c. Some samples of coal, oil, and natural gas, all supposedly many millions of years old have radiocarbon ages of less than 50,000 years.

d. Deep ocean deposits supposed to contain remains of most primitive life forms are dated within 40,000 years.

If the earth and life on earth were really as ancient as the theory of evolution requires, a great proportion of radiocarbon ages should be infinite. This is because, with a half-life of only 5730 years, initial radiocarbon in a fossil decreases in about ten half-lives to a level too low to be measured.

10. Is there scientific evidence to indicate that radiocarbon dates are in need of correction?

Answer: As we have seen, the large majority of carbon-14 ages are either within the range of biblical chronology or not far beyond it. There is evidence that radiocarbon dates should include a correction factor, and that the resulting correction would bring them into line with the Bible.

The basic assumption of the radiocarbon method is that the rate at which carbon-14 (radiocarbon) is produced in the upper atmosphere by cosmic rays has been constant for well over 50,000 years. This radiocarbon has supposedly become well mixed in the earth's circulating or exchangeable carbon supply and has built up to its maximum or equilibrium concentration. Taken in by plants and animals, it has been assumed to have been at its maximum concentration in living things throughout all of this time. Therefore, whenever a plant or animal has died and stopped taking radiocarbon into its tissues, the radiocarbon started to decrease by radioactive decomposition. Thus the amount of radiocarbon remaining in a fossil plant or animal can be measured and used as a clock to determine the time since the creature died. If the assumptions are correct, carbon-14 should provide a pretty good clock.

However, several kinds of difficulties with radiocarbon dating have come to light as experimental data accumulates. Recent research on the radiocarbon content of tree growth rings indicates that the rate of radiocarbon production has varied considerably in the past.[165] In addition some information supports the view that the amount of radiocarbon in the earth's circulating carbon supply has not yet built up to its equilibrium value. It may now only be at 70 to 80 percent of its maximum value.[166] The build-up to the maximum concentration should take only 30,000 years or less. So if radiocarbon is now only 70 to 80 percent of the way to equilibrium, this would indicate that the age of the earth's atmosphere and biosphere is only about 10,000 to 13,000 years.[167]

Thus, the assumptions involved in the carbon-14 dating method are of uncertain accuracy.

Moreover, the evidence just discussed supports the view that they are, indeed, erroneous. Therefore, it should not be surprising if the carbon-14 dating method has yielded some discordant results. That such is actually the case is attested to by the following examples:

1. Age determinations of materials from a prehistoric village site that was occupied for only about 500 years showed a spread of 6,000 years.[168]

2. The shells of living mollusks have been dated at up to 2,300 years old.[169]

3. Mortar from an English castle only 785 years old has yielded an age of 7,370 years.[170]

4. Seals freshly killed have yielded an age of 1,300 years and mummified seals dead only about 30 years were dated at 4,600 years.[171]

Facts such as those just rehearsed suggest that radiocarbon dates should be considered with caution. In particular some correction formula is needed to rectify radiocarbon ages greater than 3,500 years in order to obtain true ages.

11. Has a correction formula or curve for radiocarbon ages been worked out which brings radiocarbon dates into agreement with the Biblical time scale?

Answer: Several have been devised which look promising.[172] The principal factors which may have had important influence on the apparent radiocarbon ages which have been published include the following:

a. Initiation of cosmic radiation at time of creation only some thousands of years ago,

b. Variation of cosmic radiation from interstellar space or as a result of solar influence on the earth's magnetic field,

c. Shielding effect of a stronger pre-flood earth magnetic field,

d. Shielding effect of a pre-flood water vapor canopy,

e. Burial of much of the pre-flood exchangeable carbon at the time of the flood, and

f. Introduction of extraterrestrial radiocarbon at the time of the flood.

Different combinations of some of these factors are invoked in the several models devised for the correction of radiocarbon ages. In view of the uncertainties and greater complexities involved, all interpretations of carbon-14 data are speculative except for the period of the past 3,500 years in which calibration against historically dated materials is possible. Further research may eventually bring more general agreement among creation-science workers on a practical correction formula to bring radiocarbon dates into reasonable agreement with Biblical chronology.

N. H. Gale, J. Arden and R. Hutchinson, *Nature, Physical Science, 240* (1972), p. 56. . . . it is not widely appreciated, outside the ranks of those who work directly in geochronology or meteorites, that, judged by modern standards, the meteorite lead-lead isochron is very poorly established. . . . It therefore follows that the whole of the classical interpretation of the meteorite data is in doubt, and the radiometric estimate for the age of the Earth is placed in jeopardy.

Section K

SCIENTIFIC EVIDENCE FOR A YOUNG EARTH

1. Does human history offer any evidence for a young earth?

Answer: Yes, there was enough time from the flood for the population to grow to the estimated value at the time of Christ. But the entire solar system could not contain the population that would have developed in a million years.

Estimates of the total human population at the time of Christ center at about 300 million.[173] If the Flood was about 5000 B.C. and if the average length of a generation was forty years, Noah's family of eight people would reach 300 million by Christ's time if each family had an average of just 2.3 children. This is an average annual population increase of only 0.35 percent, whereas the present world population growth is about two percent annually. Thus the theory that the human race had been multiplying for a million years or more seems absurd, even considering the fact that modern medicine and technology were not available. For example, with an annual growth rate of only 0.01 percent, in a million years the population would be over 10^{43} people, enough to fill 3,500 solar systems solidly with bodies out to the orbit of the planet Pluto.

2. Does the earth itself offer evidence that it is young, not old?

Answer: Yes. We will list four such evidences.

a. Careful studies of the volume and rate of accumulation of the delta of the Mississippi show that it could not be older than about 5000 years.[174] This age is obtained by dividing the weight of sediments deposited annually into the total weight of the delta.

b. Petroleum and natural gas are held at high pressures in underground reservoirs of porous rock and sand. These fluids are retained in their reservoirs by relatively impermeable cap rock. However, in many cases the pressures are exceedingly high. Calculations based on the measured permeability of the cap rock show that the oil or gas pressure could not be maintained for much longer than 10,000 years or perhaps a maximum of 100,000 years. (Permeability is a measure of how easily fluids under pressure will seep through the rock.) If these fossil fuel deposits were actually millions or hundreds of millions of years old, they would long ago have leaked out through their cap rocks to the surface.[175]

c. Meteorites supposedly have been plunging to the earth's surface during the entire history of the earth, and there is no reason to believe otherwise. Therefore, if the thousands of feet of sedimentary rocks which blanket much of the earth required several billion years for their deposition, large numbers of meteorites should be embedded in them. In actual fact meteroites have been found only in the surface or younger sedimentary layers, none being discovered in the deeper, "older" strata.[176] These observations fit well with the flood geology model in which the major sedimentary strata were deposited during the creation week or during the year of the flood of Noah. In the course of the flood year only a relatively small number of meteorites could fall to be entrapped in the flood sediments. Many more meteorites would have fallen in the thousands of years since the flood, and some of these would be preserved in the recent surface sediments deposited over that period. The failure to find meteorites in deeper sediments is difficult to explain on the evolutionary model of earth history. The earth appears to be young.

d. Lord Kelvin, the eminent British physicist of the past century, was a Bible-believing Christian. He showed that if the earth were once in a molten state, from the first appearance of an initial solid crust, the time of cooling to the present temperature could not have required more than about 22 million years. More recent studies show that even taking into account the heat produced by radioactive decay in the earth's crust, the cooling time could not be more than about 45 million years.[177] This is simply not enough time for evolution to occur, in the opinion of evolutionary scientists.

3. Do the oceans speak for a young earth?

Answer: Yes. From the dissolved salts and from the sediments on the ocean floor we can conclude that the earth is young.

a. The concentrations of various elements and salts contained in sea water, when compared with the estimated annual amounts being added by rivers, subterranean springs, rain water, and other sources, uniformly point to a young age for the ocean and thus for the earth. Of fifty-one chemical elements contained in sea water, twenty could have accumulated to their present concentrations in 1,000 years or less, nine additional elements in no more than 10,000 years, and eight others in no more than 100,000 years.[178] The nitrates in the oceans could have accumulated in 13,000 years, according to a recent estimate.[179]

b. The average depth of sediments on the ocean floors is only a little more than one-half mile. If the total weight of these sediments is divided by the estimated annual addition of sediments from the continents, the age thus calculated for the oceans is only about 33 million years. This is less than one percent of the 4.5 billion year age commonly cited for the earth. In this calculation a correction has been made for the possible subduction (burial in the crust) of sediments by sliding tectonic earth plates. At present rates of erosion the continents should erode down to sea level in only about 14 million years, but there is no proof that they have yet been worn down even one time. Another way to put it is that billions of years of erosion and sedimentation should have loaded sixty miles of non-existent sediments on the ocean floors.[180] Actually, the present load of sediments was probably mostly deposited very rapidly during the period of the global flood of Noah's time.

4. Does the earth's atmosphere have anything to say about its age?

Answer: Yes. There are at least two kinds of atmospheric evidence for a young earth.

a. Helium gas resulting from radioactive decay is continually being released into the atmosphere from the earth's crust. The estimated rate of this release compared with the total helium now in the atmosphere suggests that the atmosphere may be only about 12,000 to 60,000 years old. Escape mechanisms by which helium could get away from the earth have yet to be established, and it may even be that helium from the sun is adding to the earth's atmospheric helium.[181]

b. As mentioned earlier, radioactive carbon-14 is thought to be increasing in the earth's interchangeable carbon inventory, including the atmospheric carbon dioxide. By comparing the rate of production of carbon-14 by cosmic rays with the total carbon-14 already present, the age of the atmosphere may be calculated to be only about 12,500 years.[182]

5. Does the solar system have anything to say about its age and the age of the earth?

Answer: Yes. The objects orbiting in the space between the planets suggest that the solar system and the earth are not very old.

a. Comets are loose clumps of rocky chunks, dust, and frozen gases. Each time one of them swings close to the sun it is warmed up, disturbed by the sun's gravitational force, and loses a small part of its matter. Careful analysis of the effect of this process of dissolution on the short-term comets (those returning every couple of centuries or oftener) reveals that all such comets should be totally dissipated in about 10,000 years. Since there are still many comets orbiting the sun, the solar system must not be much more than 10,000 years old. All attempts thus far to explain away this evidence for a young solar system have failed to stand critical examination.[183]

b. Meteors of all sizes down to microscopic bits of dust crash into the earth's atmosphere and settle to the surface, mostly in the form of meteroritic dust. From an estimated 14 million to as much as perhaps 50 million tons annually collect on the earth's surface.[184] If this process had been continuing for 4.5 billion years, the material would amount to a layer at least 150 feet thick, containing much higher percentages of iron and nickel than does the earth's crust in general. Such an amount is nowhere to be seen. Of course this material could have been worked into the continental crust, but on the ocean floor it should be identifiable. For example, there should be an average of over 600 pounds of nickel on each square foot of the ocean floor, but it simply is not there. Likewise, the layer many feet deep of meteoritic dust to be expected on the moon's surface has not been found. These facts point strongly to a young earth and moon.

c. Very fine dust particles orbiting the sun are pushed out into space by the pressure of solar radiation. Orbiting objects which exceed a certain minimum size, as a result of absorbing and reradiating solar energy, experience a drag effect which draws them slowly into the sun. Thus, if the solar system were just two billion years old, all objects three inches in diameter or smaller should have been swept out of space all the way to the planet Jupiter. But there are still large quantities of such materials in orbit, so the solar system must be much younger.[185]

6. Do the stars which declare the glory of God also support Biblical chronology?

Answer: Yes, a number of facts about the stars suggest a young universe.[186]

a. Some of the very bright O and B class, Wolf-Rayert, and P Cygni stars are radiating energy perhaps 100,000 to one million times as fast as our sun. They do not contain enough hydrogen to continue the necessary atomic fusion energy production at these rates for more than some tens of thousands to hundreds of thousands of years. This would suggest that perhaps the idea that stars are millions or billions of years old is wrong.[187]

b. A star cluster contains hundreds or thousands of stars moving like a swarm of bees, held together by gravity. In some clusters, however, the stars are moving so fast that the clusters could have held together only for thousands, not for millions of years.[186]

c. Binary stars (two stars orbiting around their common center of gravity) are very numerous. Many such pairs consist of two very different types of stars, one theoretically very old and the other young. How could this be if they had to evolve together in order to form a pair? Such problems have frustrated theorists in their efforts to understand how binary stars could have evolved.[188] Perhaps the great age of stars is a fiction, and binaries were created in essentially their present condition only ten thousand or so years ago along with all the other stars.

d. The galaxies are vast swarms of billions of stars interspersed with clouds of gas and dust.

They supposedly evolved from great rotating clouds of gas and dust over periods of billions of years. However, if they are that old, the spiral galaxies should have their spiral arms all twisted up, wrapped around until they have disappeared. Furthermore, the strange "barred" galaxies offer a particular problem which is poorly understood. Explanations of the physical forces which might preserve the barred structure for millions of years are highly speculative.[189] Furthermore, it has recently been concluded that the spiral galaxies appear to have the wrong amounts of random and rotational kinetic energy for stability. According to this view they all should long ago have degenerated into the barred form if they were actually billions of years old.[190]

e. Another difficult problem is posed by clusters of galaxies. The members of such a cluster are moving in different directions like a swarm of bees, and are supposedly held together by their mutual gravitational attraction. However, careful study has indicated that in some clusters there is not enough mass in all of the galaxies and the observed intergalactic matter to hold them together for millions of years.[191] In some clusters there appears to be only one-tenth to one-seventh of the required total mass. This would suggest that the clusters and their member galaxies were created rather recently and are not actually billions of years old.

7. Are scientists forced by the facts to

believe that the world and the universe are billions of years old?

Answer: No, it is their commitment to materialism and thus to evolutionary theories that forces them to adopt the great age chronologies. Materialism requires evolution to explain origins, and evolution requires long time spans in order to seem even plausible. Therefore, scientists have been looking for evidence of vast ages of time in earth history. However, the actual facts of the sciences may be interpreted within the framework of Biblical creation and a young earth and universe.

8. If the world was created in six days as the Bible reports, did not God create things with a false appearance of age? Isn't this deceptive?

Answer: The Garden of Eden was filled with false appearances of age, it is true — full-grown trees, plants, animals, an entire biosphere. In fact, Adam and Eve themselves were created as adults from the beginning of their existence. But this is not deceptive, since God has told us what He did, and we need but believe what He tells us. Those who insist that the world made itself are deceiving themselves.

Section L

THE WORLD WE LIVE IN

1. Why do some scientists say it is almost certain that life exists on planets in other solar systems and other galaxies?

Answer: This opinion is based on unproveable theories about how the earth and life supposedly evolved. These theories cannot be substantiated scientifically, and there is no evidence for the existence of extraterrestrial life.

This idea, much publicized in the mass media, hinges upon two evolutionary theories. These are the supposed evolution of the solar system from a cloud of gas and dust and the supposed evolution of life from chemicals. Neither theory can be scientifically demonstrated to be true, but if they *were* true, certain logical conclusions would follow:

a. In the universe billions of stars similar to

our sun are assumed to have evolved in the same way that our sun supposedly did. If this is true, millions of planets quite similar to the earth and having similar atmospheres, climates, etc., must likewise exist.

b. Since life supposedly evolved under these conditions on our earth, life must also have evolved on many of these other planets, developing some species as intelligent as or even more intelligent than man.

It is important to remember, however, that no scientific evidence whatsoever suggests that any other planet similar to our earth exists anywhere in the universe. The whole idea of life on other planets is pure speculation without a shred of scientific evidence.[192] But the notion is popular with both scientists and lay people who want to believe in evolution in spite of the evidence that it is impossible. God may, indeed, have created life on other planets, but He has given us no evidence of it, either through science or in the Bible.

2. Does the earth appear to be specifically designed to support human life?

Answer: Yes. The features of the earth-sun system which are essential to a life support system for plants, animals, and man are numerous.[193]

Consider these:

a. The sun's temperature is right to provide the range of light wavelengths suitable for life.

Higher temperatures would result in too much ultraviolet radiation, lower temperatures in too much infrared.

b. The sun is at the correct distance from the earth and has the proper size and temperature to provide the total amount of radiation required to maintain surface temperatures on earth suitable for life. Appreciable changes in any of these factors would probably destroy all life.

c. The nearly circular orbit of the earth limits temperature variations.

d. The water vapor and carbon dioxide in the atmosphere produce a so-called "greenhouse effect" which moderates the temperature extremes.

e. A high altitude ozone layer effectively absorbs the lethal fraction of solar ultraviolet rays which would destroy life on the earth's surface were the ozone layer removed.

f. The 23½ degree inclination of the earth's axis of rotation from the perpendicular to the plane of its orbit provides for the seasons. It probably also considerably increases the land area in the northern hemisphere suitable for intensive summer agricultures.

g. The earth's magnetic field, extending tens of thousands of miles into space, shields the earth's surface from much of the cosmic radiation which probably would prove deleterious to life.

h. The lunar gravitation produces important tidal circulation effects in the oceans, which make conditions much more suitable for sea life in the shallow zones along shores and in estuaries.

i. The mass and size of the Earth are adjusted to provide gravitational force and atmospheric pressure suitable for life.

j. The two major constituents of the atmosphere. oxygen (21%) and nitrogen (78%), are balanced to make up the ideal medium for the support of life.

k. The earth's surface is blessed with a concentration unknown anywhere else in the universe, of liquid water, the only possible solvent and medium for living cells. The physical properties of water are unique and absolutely essential to all life processes. These properties of water include (1) the highest heat of fusion (melting), (2) a liquid temperature range which includes the temperatures at which enzymes and other life molecules can exist and function, (3) the highest heat capacity (heat required to heat it up), (4) the highest heat of vaporization, (5) and the highest dielectric constant of all of the common liquids; (6) the greatest solvent powers, (7) the greatest power to form a special chemical bond called a hydrogen bond, and (8) the greatest ionizing power of all the common liquids; (9) the property (unique except for bismuth) of expanding when it crystallizes, and (10) a strong power to absorb infared radiation.

Liquid ammonia is sometimes proposed as a possible liquid medium for some imagined kind of life system. It has slightly higher values in several of the above properties, but it is distinctly inferior to water in the others. The idea of life in any solvent other than water is mere fantasy.

1. The elements carbon, hydrogen, oxygen, nitrogen, and phosphorus, together with liquid water are the basis for the only remotely possible chemical framework for life. The surface of the earth is the only known place in the entire universe where these substances are found in suitable quantities and in the proper forms for life to exist.

The ideal combination of conditions and factors essential to life observed only on earth surely points to intelligent, purposeful design. To believe that this beautifully balanced life support system which carries the human race safely at speeds of more than one thousand miles per minute through hostile space is a mere accident requires invincible faith in the power of chance. Would not any honest observer have to admit that the earth appears to have been designed for us to live on, just as David the Psalmist said? "The heaven, even the heavens, are the LORD'S: but the earth hath he given to the children of men." Psalm 115:16

APPENDIX 1
Mathematical Probability for Chance Beginning of Life

As explained in Section C-2, the mathematical theory of probability teamed up with sufficient time seems at first glance to favor chance evolution of life. We will now show that proper application of probability theory demonstrates the utter impossibility of the chance chemical formation of life. First, let us explain the mathematics involved.

Consider the probability of flipping a penny ten times and getting ten heads. The probability of getting a head on the first flip is $\frac{1}{2}$; on the first two flips, $\frac{1}{2} \times \frac{1}{2} = \frac{1}{4}$; with three flips, $\frac{1}{2} \times \frac{1}{2} \times \frac{1}{2} = \frac{1}{8}$; and so on, until for ten flips the probability of getting all heads is $\frac{1}{2}$ multiplied by itself ten times, or $(\frac{1}{2})^{10} = 1/1024$. This is a rather improbable event, ten heads in a row. How many times would we have to repeat the ten-flip trial to reach some reasonable probability of having at least one success, ten heads in ten flips? We will simply state here the necessary formula which is developed in detail in our book, *The Creation Explanation*.[24] If the probability of success in one trial is P(1), and if the number of trials performed is N, then the probability of at least one success in N trials is $P(N) = 1\text{-}[1\text{-}P(1)]^N$

So for our ten-fold penny-flipping experiment P(1) = 1/1024, and let us repeat the experiment 100 times. Then $P(100) = 1 - (1\text{-}1/1024)^{100} =$ 0.093 or roughly one chance in eleven, which is much better than one chance out of 1024. If we repeat the experiment 1000 times, this probability of at least one success of ten heads in a row increases to 0.624 or roughly two chances out of three. Thus from pure mathematics it appears that the evolutionist claim might be true after all. Perhaps with billions of years for chance chemical reactions, the very improbable chemical beginning of some form of life would become practically inevitable. But in order to test this mathematic probability argument, it is necessary to apply it to a world at least somewhat similar to the real one. Two things must be done. First, a reasonable estimate must be made of the actual numerical value of P(1), the probability of success in one trial. Also, N, the number of trials that would have been possible must be estimated.

Let us make the ridiculously optimistic assumption that the original life form could begin with just one particular simple enzyme molecule with only one hundred amino acid units in its chain. (A protein molececule is a long chain of various of the the twenty common amino acid molecules linked together. The order in which the different amino acids appear in the chain determines what kind of a protein it is. An enzyme is a protein designed to catalyze or promote a particular chemical reaction in a living organism.) It can be shown[24] that, with other assumptions heavily weighted on the side of success, the probability of successful formation of this enzyme molecule by one trial combination of one hundred amino acid molecules in an ancient ocean would be a very small fraction, $1/10^{60} = 10^{-60}$. This is the fraction with the number one in the numerator and the number one followed by sixty zeros in the denominator. So $P(1) = 10^{-60}$.

Now if we assume that one percent of the nitrogen in the earth's atmosphere were converted into 100-unit enzyme molecules, that would total about 10^{40} molecules, which if spread out evenly would make a layer of about 95 pounds of enzyme molecules per square foot over the entire surface of the earth. And if we assume this happened annually for one billion years (i.e., 10^9 years), the total number of trial formations of 100-unit enzyme molecules would be $10^{40} \times 10^9 = 10^{49}$ (i.e., the number one followed by 49 zeros.) Therefore, we take for our total number of trials $N = 10^{49}$. If we substitute these two numbers into our formula we get $P(10^{49}) = 1-(1-10^{-60})^{10^{49}} =$

$10^{-11} = 1/10^{11}$, or one chance in 100 billion.[24]

This means that in order to have one chance in 100 billion of finding at least one of the enzyme molecules necessary to start life, we would have to search molecule by molecule through 260 trillion tons of protein muck every year for one billion years. But it is quite certain that life could not start with just one enzyme molecule, so this calculated probability is far too high. See Section C-2 for the completion of this discussion.

Haskings, Caryl P., *American Scientist, 59,* May-June 1971, p. 305. . . . But the most sweeping evolutionary questions at the level of biochemical genetics are still unanswered. . . . The fact that in all organisms living today the processes both of replication of the DNA and of the effective translation of its code require highly precise enzymes and that, at the same time, the molecular structures of those same enzymes are precisely specified by the DNA itself, poses a remarkable evolutionary mystery.

Did the code and the means of translating it appear simultaneously in evolution? It seems almost incredible that any such coincidence could have occurred, given the extraordinary complexities of both sides and the requirement that they be coordinated accurately for survival. By a pre-Darwinian (or a skeptic of evolution after Darwin) this puzzle surely would have been interpreted as the most powerful sort of evidence for special creation.

Dixon, M. and Webb, E., *Enzymes* (New York: Academic Press, 1964), p. 670.

Thus the whole subject of the origin of enzymes, like that of the origin of life, which is essentially the same thing, bristles with difficulties. We may surely say of the advent of enzymes, as Hopkins said of the advent of life, that it was 'the most improbable and the most significant event in the history of the Universe.'

APPENDIX 2
What the Bible Is All About and the Reason for *Handy-Dandy*

The reason God the Creator worked for some fifteen hundred years through almost forty prophets and apostles to bring us the Bible is given in a verse in the New Testament. John 20:31 says, "But these have been written that you may believe that Jesus is the Christ, the Son of God; and that believing you may have life in His name." So the Bible is really about Jesus Christ, the eternal Son of God, what He did to save us from our sins, and how we can have forgiveness and eternal life by faith in the Lord Jesus.

But to have faith in Jesus Christ and be saved a sinner must believe what the Bible says about his personal sin and guilt before a holy God and about what Christ has done to save him. Anything, therefore, which stands in the way of faith in the Bible as the Word of God can keep sinful men and women from the Savior Whom they must know or perish. Supposedly scientific theories such as evolution which contradict the Bible can cause some people to doubt the Bible and thus hinder them from coming in humble faith to Jesus Christ for salvation.

The *Handy-Dandy Evolution Refuter* provides logic and scientific evidence to show that materialistic evolutionary theories are really not science, and that there is actually no scientifically based reason for ignoring or refusing the gracious offer of God to save those who will believe in His Son Jesus Christ. It is our hope that our readers will come to faith or to stronger faith in the Bible and in the God of the Bible Who is Creator, Lord, and Judge of the world.

References

Section A

1 Popper, Karl, *The Logic of Scientific Discovery,* Basic Books, Inc., New York (1959), pp. 40-42.

2 Matthews, L. Harrison, Introduction to *The Origin of Species* by Charles Darwin, J.M. Dent & Sons Ltd., London (1971), p. xii.

3 Birch, L.C. and Ehrlich, P.R., *Nature, 214,* 22 April 1967, p. 352; Popper, Karl R., *Federation Proceedings, Amer. Societies for Experimental Biology, 22* (1963) p. 964; Olson, Everett C., *Evolution After Darwin, Vol. 1,* Sol Tax, editor, Univ. of Chicago Press (1960), pp. 530-537.

4 John 1:1-3, Colossians 1:16, Hebrews 1:1-3, John 5:45-47, Matthew 19:4-6.

5 Thompson, W.R., Introduction to *The Origin of Species* by Charles Darwin, E.P. Dutton and Co., New York (1956).

6 MacDonald, D.K.C., *Faraday, Maxwell, and Kelvin,* Doubleday and Co., Inc., Garden City, N.Y. (1964), pp. 10-11; Tyndal, John, *Faraday as a Discoverer,* Thomas Y. Crowell Co., New York (1961), pp. 43-48, 176-199.

7 MacDonald, D.K.C., *op. cit.* (ref. 6), pp. 63, 97-98; Watson, Dwight, *The Scientists: James Clerk Maxwell and Michael Faraday,* Bible-Science Association, Caldwell, Idaho (1973).

8 MacDonald, D.K.C., *op. cit.* (ref. 6), p. 63.

9 The Creation Research Society with membership office at 2717 Cranbrook Road, Ann Arbor, Mich. 48104, is a professional society with some 450 voting members having advanced degrees in science who believe the biblical record of creation and the flood.

Section B

10 Simpson, G.G., *The Meaning of Evolution,* Bantam Books, Inc., New York (1971), pp. 314-315.

11 Zeiller, Warren, *Natural History,* Dec. 1971, pp. 36-41.

12 Odum, Eugene P., *Fundamentals of Ecology,* 3rd Edition, W.B. Saunders Co., Philadelphia (1971), pp. 273-274; Janzen, Daniel H., *Science, 188,* 30 May 1975, pp. 936-937.

13 Sauer, E.G.F., *Scientific American, 199,* Aug. 1958, p. 42; Emlen, Stephen T., *ibid., 233,* Aug. 1975, pp. 102-111.

14 Keeton, William T., *ibid, 231,* Dec. 1974, pp. 96-107.

15 Kaufmann, David A., *A Challenge to Education II-A,* Walter Lang, Editor, Bible-Science Association, Inc.,

Caldwell, Idaho (1974), pp. 119-130.

Section C

16 Miller, Stanley L. and Orgel, Leslie E., *The Origins of Life On the Earth,* Prentice-Hall, Inc., New Jersey (1974), p. 59.
17 *Ibid.,* p. 33; Abelson, P.H., *Proceedings, National Academy of Science, 55,* 1966, pp. 1365-1372.
18 Brinkman, R.T., *Journal of Geophysical Research, 74,* 1969, p. 5335.
19 Ferris, J.P., and Nicodem, D.E., *Nature, 238,* 1972, pp. 268-269.
20 Miller, Stanley L. and Orgel, Leslie E., *op. cit.* (ref. 16), pp. 83-117.
21 *Ibid.,* pp. 118-128, 135-151.
22 Watson, James D., *Molecular Biology of the Gene,* 2nd Edition, W.A. Benjamin, Inc., New York (1970), p. 149.
23 Paecht-Horowitz, M., *et al., Nature, 228,* 14 Nov. 1970, p. 636.
24 Kofahl, Robert E. and Segraves, Kelly L., *The Creation Explanation,* Harold Shaw Publishers, Wheaton, Ill. (1975), pp. 98-100, 239.
25 Kenyon, D.H. and Steinman, G.D., *Biochemical Predestination,* McGraw-Hill, New York (1969).
26 Miller, Stanley L., and Orgel, Leslie E., *op. cit.* (ref. 16), p. 144.
27 Morowitz, H.J., *Progress in Theoretical Biology, 1* (1967), pp. 50-58; Coppedge, James F., *Evolution: Possible or Impossible?,* Zondervan Publishing House, Grand Rapids (1973), p. 110.
28 Morowitz, H.J., *Energy Flow in Biology,* Academic Press, New York (1968), p. 99.
29 Miller, Stanley L. and Orgel, Leslie E., *op. cit.* (ref. 16), p. 164.

Section D

30 Morris, Henry M., Editor, *Scientific Creationism,* Creation-Life Publishers, San Diego (1974), pp. 38-46; Wysong, R.L., *The Creation-Evolution Controversy,* Inquiry Press, East Lansing, Mich. (1976), pp. 239-263; Kofahl, Robert E., and Segraves, Kelly S., *op. cit.* (ref. 24), pp. 21-38.
31 Morowitz, H.J., *op. cit.* (ref. 28), pp. 21-43; Groth, W., *Photochemistry in the Liquid and Solid States,* F. Daniels, Editor, John Wiley & Sons, Inc., New York (1960), p. 21

32 Watson, James D., *op. cit.* (ref. 22), pp. 268, 269, 302.
33 Huxley, Julian, *Evolution in Action,* Harper Bros., New York (1953), p. 41.
34 Simpson, G.G., *Biology and Man,* Harcourt, Brace & World, New York (1969), p. 127; Williams, George C., *Adaptation and Natural Selection,* Princeton Univ. Press (1966), pp. 54, 139.
35 Williams, Emmett L., Jr., *Creation Research Soc. Quarterly, 8,* Sept. 1971, pp. 117-126.

Section E

36 Wills, Christopher, *Scientific American, 222,* March 1970, p. 98.
37 Crow, James F., *Bulletin of Atomic Scientists, 14,* Jan. 1958, pp. 19-20, quoted in *The Genesis Flood* by Whitcomb, John C., Jr., and Morris, Henry M., p. 401.
38 Klotz, John W., *Genes, Genesis, and Evolution,* Concordia Publishing House, Saint Louis (1970), pp. 262-265.
39 Moore, Jerry P., *Creation Research Soc. Quarterly, 10,* March 1974, pp. 187-190.
40 Weinberg, Janet H., *Science News, 107,* 22 Feb. 1975, pp. 124-127.
41 Klotz, John W., *op. cit.* (ref. 38), pp. 291-313.
42 Campbell, J.H., *et al., The Proceedings of the National Academy of Science U.S., 70* (1973), pp. 1841-1845.
43 Mayr, Ernst, *Population, Species, and Evolution,* Harvard Univ. Press, Cambridge, Mass. (1970). The variations cited in Mayr's work are of this limited type.
44 Bishop, A.J. and Cook, L.M., *Scientific American, 232,* Jan. 1975, pp. 90-99.
45 Matthews, L. Harrison, *op. cit.* (ref. 2), p. xi.

Section F

46 Romer, Alfred, *Vertebrate Paleontology,* 3rd Edition, Univ. of Chicago Press (1966), p. 314.
47 Simpson, G.G., *The Meaning of Evolution,* Bantam Books, Inc., New York (1971), pp. 16-19; Axelrod, Daniel, I., *Science, 128,* 4 July 1958, p. 7.
48 Romer, Alfred, *op. cit.* (ref. 46), pp. 15, 23, 25, 47; Ommanney, F.D., *The Fishes,* Time Inc., New York (1964), p. 60; *The Fossil Record,* Geological Society of London (1967), p. 630.
49 *Ibid.,* p. 79.
50 *Ibid.,* pp. 86-91.

51 Gish, Duane, *Evolution, the Fossils Say No!,* Creation-Life Press, San Diego (1973), pp. 53-57.
52 *The Fossil Record,* Geological Society of London (1967), pp. 686, 696; Gish, Duane, *op. cit.* (ref. 51), pp. 58-59; Romer, Alfred, *op. cit.* (ref. 46), p. 173.
53 Romer, Alfred, *op. cit.* (ref. 46), pp. 95, 102.
54 Gish, Duane, *op. cit.* (ref. 51), p. 58; Romer, Alfred, *op. cit.* (ref. 46), pp. 104, 187-88, 191; *McGraw-Hill Encyclopedia of Science and Technology,* New York (1971), Vol. 4, pp. 359-360.
55 Goldschmidt, Richard, *The Material Basis of Evolution,* Pageant Books, Inc., Paterson, N.J. (1960), pp. 6-7.
56 Romer, Alfred, *op. cit.* (ref. 46), pp. 166-167, 368, 374.
57 Wysong, R.L., *op. cit.* (ref. 25), pp. 300-301.
58 Romer, Alfred, *op. cit.* (ref. 46), p. 140.
59 Schmidt-Nielsen, Knut, *Scientific American, 225,* Dec. 1971, pp. 72-79.
60 *The Fossil Record* (ref. 52), pp. 107, 117, 508.
61 Romer, Alfred, *op. cit.* (ref. 46), pp. 144-147.
62 *Ibid.,* p. 338; Jepsen, G.L., *Science, 154,* 9 Dec. 1966, p. 1333.
63 Simpson, *G.G., Evolution After Darwin,* Vol. 1, Sol Tax, Editor, University of Chicago Press (1960), p. 143; also ref. 47, p. 209; Kitts, David B., *Evolution, 28,* Sept. 1974, p. 467.
64 Simpson, G. G., *Temp and Mode in Evolution,* Columbia Univ. Press, New York (1944), pp. 105-108; White, Errol, *Proc. Linnean Soc. London, 177,* Jan. 1966, p. 8; George, Neville T., *Science Progress, 48,* Jan. 1960, p. 3.
65 Nilsson, Heribert, *Synthetische Artbildung,* Verlag C.W.K. Gleerup, Lund, Sweden (1953), reprint of English summary published by Evolution Protest Movement of North America, Victoria, B.C. (1953), pp. 1211-1212; Corner, E.J.H., *Evolution in Contemporary Botanical Thought,* A.M. MacLeod and L.S. Cobley, Editors, Quadrangle Books, Chicago (1961); Arnold, C.A., *An Introduction to Paleobotany,* McGraw-Hill, New York (1947), p. 7.
66 Davidheiser, Bolton, *Evolution and Christian Faith,* Presbyterian and Reformed Pub. Co., Nutley, N.J. (1969), pp. 303-313.
67 Simpson, G.G., *op. cit.* (ref. 64), p. 167; Cousins, Frank W., *Creation Research Soc. Quarterly, 8,* Sept. 1971, pp. 99-108; Nilsson, Heribert, *op. cit.* (ref. 65), pp. 1193-1194; Kerkut,

G.A., *Implications of Evolution,* Pergamon Press, New York (1960), pp. 144-149; Wentworth, Baroness, *Thoroughbred Racing Stock,* Charles Scribners, Sons, New York (1938), p. 379.
68 Simpson, G.G., *Horses,* Oxford Univ. Press, New York (1951), pp. 105-112, 115-116.
69 *Ibid.*, pp. 116-117, 121; Simpson, G.G., *op. cit.* (ref. 10), p. 135.
70 *Ibid.*, p. 124. Other fossil horse data cited below can be found in the same work.
71 *Ibid.*, pp. 177-179; Davidheiser, Bolton, *Creation Research Soc. Quarterly, 12,* Sept. 1975, pp. 88-89.
72 Kurten, Bjorn, *The Age of the Dinosaurs,* McGraw-Hill Book Co., New York (1968), pp. 234-235.
73 *Footprints in Stone,* Films for Christ Association, R.R. 2, Eden Road, Elmwood, Ill. 61529.
74 Wysong, R.L., *op. cit.* (ref. 30) pp. 287-288.
75 Jepsen, G.L., *Science, 154,* 9 Dec. 1966, p. 1333.
76 *The Fossil Record,* (ref. 52), Sphenodontia, p. 702; Bogert, C.M., *Scientific Monthly, 76,* March 1953, p. 165.
77 *Ibid.*, (ref. 52), Tryblidioidea, p. 423; Hickman, C.P., *Integrated Principles of Zoology,* 2nd Edition, The C.V. Mosby Co., St. Louis (1961), pp. 348-349.
78 *Ibid.*, (ref. 52), Blattodea, p. 510; Brues, C.T., *Scientific American, 185,* Nov. 1951, p. 57.
79 *Ibid.* (ref. 52), Odonata, pp. 510-511; *Science Digest, 49,* Jan. 1961, p. 6.
80 *Ibid.* (ref. 52), Asteroidea, pp. 592-595; Nelson, Byron, *After Its Kind,* Bethany Press, Minneapolis (1967), quote from Smithsonian Institute Bulletin 88.
81 Chaney, R., *American Scientist, 36,* Oct. 1948, p. 490.
82 *The Fossil Record* (ref. 52), pp. 260-261; Delevoryas, T., *Morphology and the Evolution of Fossil Plants,* Holt, Rinehart and Winston, New York (1962), pp. 164-165.
83 *Ibid.* (ref. 52), Cycadales, pp. 261, 263; Stokes, W.L., *Essentials of Earth History,* Prentice-Hall, New Jersey (1960), p. 266.
84 Romer, Alfred, *op. cit.* (ref. 46), pp. 74-75; *Life,* 3 April 1939, p. 26.
85 *The Fossil Record* (ref. 52), p. 41.
86 *Ibid.* (ref. 52), Articulata, pp. 574-575; Idyll, C., *Abyss—the Deep Sea and the Creatures that Live In It,* Crowell Pub. Co., New York (1971), pp. 237-238.

87 *Ibid,* (ref. 52) Echinoidea, pp. 586-589.
88 *Ibid.* (ref. 52) Teuthoidea, pp. 463; Idyll, C., *op. cit.* (ref. 86), pp. 251-253.

Section G

89 Weiner, S.J., *The Piltdown Hoax,* Oxford Univ. Press (1955).
90 Davidheiser, Bolton, *op. cit.* (ref. 66), pp. 330-334.
91 Straus, Wm. L., Jr. and Cave, A.J.E., *Quarterly Review of Biology, 32,* Dec. 1957, pp. 348-363.
92 Cousins, Frank W., *Fossil Man,* Evolution Protest Movement, Hants, England (1971), pp. 40-42.
93 O'Connell, Patrick, *Science of Today and the Problems of Genesis,* Christian Book Club of America, Hawthorne, Calif. (1969), pp. 139-142.
94 *Ibid.,* pp. 108-138.
95 Leakey, R.E.F., *Nature, 231,* 28 May 1971, pp. 241-245.
96 Oxnard, C.E., *Nature,* 258, 4 Dec. 1975, pp. 389-395; Zuckerman, Sir Solly, *Journal of the Royal College of Surgeons of Edinburgh* (1966), Vol. II (2), pp. 87-114; *Beyond the Ivory Tower,* Taplinger Pub. Co., N.Y. (1971), pp. 76-94.
97 Leakey, Richard E., *National Geographic Magazine, 143,* June 1973, p. 819.
98 Taieb, Maurice, and Johanson, Karl, quoted in *Scientific American, 231,* Dec. 1974, p. 64; *Science News, 109,* 13 March 1976, pp. 164-165.
99 Smith, John Maynard, *On Evolution,* Edinburgh Univ. Press (1972), pp. 109-112; Calder, Nigel, *The Life Game,* Viking Press, Inc., N.Y. (1974), pp. 24, 68-70; Bass, Robert W., *Creation Research Soc. Quarterly, 12,* March 1976, p. 199.
100 McCone, R. Clyde, *Symposium on Creation IV,* Donald W. Patten, Editor, Baker Book House, Grand Rapids, Mich. (1972), pp. 123-133.
101 *Time,* 30 June 1967, p. 34.
102 Koontz, Robert F., *Creation Research Soc. Quarterly, 8,* Sept. 1971, pp. 128-129.

Section H

103 Whitcomb, John C., Jr. and Morris, Henry M., *The Genesis Flood,* Presbyterian and Reformed Pub. Co., Philadelphia (1961), pp. 84, 128-130, 155-157.
104 Velikovsky, Immanuel, *Earth In Upheaval,* Dell Pub. Co., Inc., New York (1955), pp. 50-60; Whitcomb, John C., Jr. and Morris, Henry M., *op. cit.* (ref. 103), pp. 154-161.

105 Nilsson, Heribert, *op. cit.* (ref. 65), pp. 1194-1196.
106 Velikovsky, Immanuel, *op. cit.* (ref. 104), pp. 18-22, 64-69, 78-81.
107 Nilsson, Heribert, *op. cit.* (ref. 65), pp. 1196-1198; Morris, Henry M., Editor, *op. cit.* (ref. 30), pp. 107-109.
108 Rupke, N.A., *Creation Research Soc. Quarterly, 3,* May 1966, pp. 16-37.
109 Whitcomb, John C., Jr. and Morris, Henry M., *op. cit.* (ref. 103), pp. 278-279.
110 Coffin, Harold, *A Challenge to Education II-B,* Walter Lang, Editor, Bible-Science Association, Caldwell, Idaho (1974), pp. 36-43.
111 Nevins, Stuart E., *A Symposium on Creation III,* Donald W. Patten, Editor, Baker Book House, Grand Rapids, Mich. (1971), pp. 33-68.
112 Wheeler, Harry E., *Bull, of the Amer. Assoc. of Petroleum Geologists, 47,* p. 1507.
113 Stokes, W.L., *Bull of the Geolog. Soc. of Amer., 61* (1950), p. 91.
114 Nevins, Stuart, *op. cit.* (ref. 111), pp. 59-60.
115 Jopling, Alan V., *Journal of Sedimentary Petrology, 36* (1966), p. 883.
116 McKee, Edwin D., *Primary Sedimentary Structures and Their Hydrodynamic Interpretation,* Gerald V. Middleton, Editor, Soc. of Economic Paleontologists and Mineralogists, Tulsa, Okla. (1965), p. 83.
117 Kuenen, Ph.H. and Migliorini, C.I., *Journal of Geology, 58* (1950), pp. 91-127.
118 Morris, Henry M., Editor, *op. cit.* (ref. 103), p. 104.
119 Dunbar, C.O., and Rogers, John, *Principles of Stratigraphy,* John Wiley & Sons, Inc., New York (1957), p. 237.
120 *Ibid.,* p. 245.
121 Pettijohn, F.G., *Sedimentary Rocks,* 2nd edition, Harper and Row, New York (1957), p. 313.
122 *Ibid.,* 2nd edition, p. 483.
123 Baker, C.L., *Journal of Geology, 31* (1923), pp. 66-79; Wadia, D.N., *Geology of India,* 3rd edition, MacMillan & Co., London (1953), pp. 291-292.
124 Daly, Reginald, *Earth's Most Challenging Mysteries,* Craig Press, Nutley, N.J. (1972), pp. 242-243; Whitcomb, John C., and Morris, Henry M., *op. cit.* (ref. 103), 125-126, 324-325.

125 Velikovsky, Immanuel, *op. cit.* (ref. 104), pp. 81-87, 151.
126 Read, John G., *The Mountains of Ararat,* filmstrip by Creation-Science Research Center, San Diego (1973).
127 Burdick, Clifford L., *Creation Research Soc. Quarterly, 6,* Sept. 1969, pp. 96-106; *ibid., 11,* June 1974, pp. 56-60; Slusher, Harold S., *ibid., 3,* May 1966, pp. 59-60; Read, John G., *Fossils, Strata, and Evolution,* filmstrip by Scientific-Technical Presentations, Culver City, Calif.
128 Burdick, Clifford, *Creation Research Soc. Quarterly, 9,* June 1972, p. 25; Axelrod, Daniel, *Evolution, 13,* 1959, pp. 264-275; Leclerque, S., *ibid., 10,* 1956, pp. 109-113.
129 Kofahl, Robert E., and Segraves, Kelly L., *op. cit.* (ref. 24), pp. 221-229; Morris, Henry M., Editor, *op. cit.* (ref. 103), pp. 203-215.

Section I

130 Klotz, John W., *op. cit.* (ref. 38), pp. 128-131.
131 *Ibid.*, pp. 120-128.
132 Davidheiser, Bolton, *op. cit.* (ref. 66), pp. 261-262.
133 Mayr, Ernst, *Systematics and the Origin of Species,* Columbia Univ. Press, New York (1942), p. 4; Davidheiser, Bolton, *op. cit.* (ref. 66), pp. 255-267.
134 Klotz, John W., *op. cit.* (ref. 38), pp. 131-136.
135 de Beer, Gavin, *Embryos and Amcestors,* rev. ed., Oxford Univ. Press (1954), pp. 6, 10; Leach, James W., *Functional Anatomy—Mammalian and Comparative,* McGraw-Hill Pub. Co., New York (1961); Davidheiser, Bolton, *op. cit.* (ref. 66), pp. 240-254; *Evolution—Science Falsely So-Called,* 19th Edition, International Christian Crusade, Toronto (1974), pp. 20-25.
136 Klotz, John W., *op. cit.* (ref. 38), pp. 145-154; P.R. Ehrlich and R.W. Holm, *The Process of Evolution,* McGraw-Hill Pub. Co., New York (1963), p. 66.
137 Dickerson, Richard E., *Scientific American, 226,* April 1972, pp. 58-72.
138 Kofahl, Robert E., and Segraves, Kelly L., *op. cit.* (ref. 24), pp. 165-169.
139 Crowson, R.A., *Nature, 254,* April 3, 1975, p. 464.
140 Davidheiser, Bolton, *op. cit.* (ref. 66), 88-105.

Section J

141 Kofahl, Robert E., and Segraves, Kelly L., *op. cit.* (ref. 24), pp. 181-213; Morris, Henry M., Editor, *op. cit.* (ref. 96), pp. 137-149; Wysong, R.L., *op. cit.* (ref. 30), pp. 145-158.

142 Cook, Melvin A., *Prehistory and Earth Models,* Max Parish, London (1966), pp. 1-89.
143 Anderson, J.L., and Spangler, G.W., *Journal of Physical Chemistry, 77* (1973), p. 3114; *Bull. of the Amer. Physical Soc., 10* (1971), p. 1180; *Pensee, 4,* Fall, 1974, pp. 31-32.
144 Jueneman, F.B., *Industrial Research, 14* (1972), p. 15; Cook, Melvin A., *op. cit.* (ref. 136), pp. 41-62.
145 York, D., and Farquhar, R.M., *The Earth's Age and Geochronology,* Pergamon Press, New York (1972), pp. 75-92; Hamilton, E.I., *Applied Geochronology,* Academic Press, New York (1965), pp. 142-149.
146 Morris, Henry M., Editor, *op. cit.* (ref. 96), pp. 140-149.
147 Driscoll, Evelyn, *Science News, 101,* 1 Jan. 1972, p. 12.
148 Cook, Melvin, *op. cit.* (ref. 142), pp. 53-62.
149 McDougall, I., Report: The *Present Status of the Decay Constants,* Subcommission on Geochronology, Bern, Switzerland, 19 Sept. 1974, p. 3.
150 Rancitelli, L.A., and Fisher, D.E., *Planetary Science Abstracts,* American Geophysical Union (1967), p. 154.
151 Armstrong, Richard Lee, "Proposal for Simultaneous Adoption of New U, Th, Rb, and K Decay Constants for Calculation of Radiometric Dates," unpublished paper, Dept. of Geological Sciences, Univ. of British Columbia, 1975; "Report on Decay Constants," 10 May 1975.
152 Cook, Melvin, *op. cit.* (ref. 142), pp. 65-66.
153 Funkhouser, J.G., and Naughton, J.J., *Journal of Geophysical Research, 73,* 15 July 1968, p. 4601; Laughlin, A.W., *ibid., 74,* 15 Dec. 1969, pp. 6684-6689.
154 Engels, Joan C., *Journal of Geology, 79,* Sept. 1971, p. 609.
155 For other examples see Kofahl, Robert E., and Segraves, Kelly L., *op. cit.* (ref. 24), pp. 200-202.
156 Obersby, V.M. *Geochimica et Cosmochimica Acta, 36,* Oct. 1972, p. 1167.
157 Wang, R.K., *et al., Science, 167,* 30 Jan. 1970, pp. 479-480.
158 Tera, F., *et al., Earth and Planetary Science Letters, 14* (1972), p. 281-303.
159 Nunes, P.D., and Tatsumoto, M., *Science, 182,* 30 Nov. 1973, p. 916.
160 Zartman, *et al., Science, 145,* 31 July 1964, pp. 479-481.
161 Cherdyntsev, V.V., *et al.,* Geolog. Institute Academy of Sciences, USSR, Earth Science Section, *172,* p. 178. The

data is reproduced by Sidney P. Clementson in *Creation Research Soc. Quarterly, 7,* Dec. 1970, p. 140.
162 Schindewolf, O., *American Journal of Science, 255,* June 1957, p. 394.
163 Wasserburg, G.J., and Lanphere, M.A., *Geological Soc. of America Bulletin, 76* (1965), pp. 735-758.
164 Whitelaw, R.L., *Creation Research Soc. Quarterly, 7,* June 1970, pp. 56-71, 83.
165 Renfrew, Colin, *Scientific American, 225,* Oct. 1971, p. 67.
166 Cook, Melvin A., *op. cit.* (ref. 142), pp. 1-10; Hess, W.M., *et al., Journal of Geophysical Research, 66* (1961), p. 665; Lingenfelter, R.E., *Review of Geophysics* (1963), p. 51; Suess, H.E., Journal of Geophysical Research, (1965), p. 5946.
167 Cook, Melvin A., *op. cit.* (ref. 142), pp. 1-10.
168 Reed, C.A., *Science, 130,* 11 Dec. 1959, p. 1630.
169 Kieth, M., and Anderson, G., *Science, 141,* 16 Aug. 1963, p. 634.
170 Baxter, M.S., and Walton, A., *Nature, 225,* 7 March 1970, pp. 937-938.
171 Dort, W., *Antarctic Journal of the U.S., 6* (1971), p. 210.
172 Whitelaw, R.L., *loc. cit.* (ref. 156); Clementson, Sidney P., *Creation Research Soc. Quarterly, 10,* March 1974, pp. 229-236; Camping, Harold, *Adam When?,* Frontiers for Christ, Alameda, Calif. (1974), pp. 178-229; Cook, Melvin A., *op. cit.* (ref. 125), pp. 1-10; Brown, R.H., *Origins, 3,* No. 2 (1976).

Section K

173 Coale, Ansley J., *Scoentific American, 231,* Sept. 1974, p. 43.
174 Allen, Benjamin F., *Creation Research Soc. Quarterly, 9,* Sept. 1972, pp. 96-114.
175 Cook, Melvin A., *op. cit.* (ref. 142), pp. 254-262; Dickey, P., *et al., Science, 160,* 10 May 1968, p. 609.
176 Heide, Fritz, *Meteorites,* Edward Anders and Eugene DuFresne, translators, Univ. of Chicago Press (1964), p. 119ff.; Nininger, H.H., in *The Moon, Meteorites, and Comets,* Barbara M. Middlehurst and Gerard P. Kuiper, editors. Univ. of Chicago Press (1963), p. 164; Steveson, Peter A., *Creation Research Soc. Quarterly, 12,* June 1975, pp. 23-25.
177 Ingersoll, Zobel, and Ingersoll, *Heat Conduction With Engineering, Geological, and Other Applications,* Univ. of Wisc. Press (1954), pp. 99-107.

178 Riley, J.B., and Skirrow, G., Editors, *Chemical Oceanography,* Vol. 1, Academic Press, London (1965), pp. 164-165.
179 Martin, Dean F., *Marine Chemistry,* Vol. 2, Marcel Dekker, Inc., New York (1970), pp. 228-263.
180 Nevins, S.E., Creation—*Acts, Facts, Impacts,* Institute for Creation Research Pub. Co., San Diego (1974), p. 164.
181 Cook, Melvin A., *op. cit.* (ref. 142), pp. 10-14; also by same author, "Age" and Rare Gas Content of Lunar Rocks and Soil", ca. 1972.
182 *Ibid.,* p. 6.
183 Lyttleton, R.A., *Mysteries of the Solar System,* Oxford Clarendon Press (1968), p. 147; Joss, P.C., *Astronomy and Astrophysics, 25,* No. 2, pp. 271-273.
184 Pettersson, Hans, *Scientific American, 202,* Feb. 1960, p. 132; Hawkins, G.S., *Annual Review Astronomy and Astrophysics, 2,* pp. 140-164, pertinent graph reproduced by Kaula, W.M., in *An Introduction to Planetary Physics,* John Wiley & Sons, Inc., New York (1968), p. 249.
185 Slusher, Harold, *Science and Scripture,* Sept.-Oct. 1971, p. 26; Abell, George, *Exploration of the Universe,* Holt, Rinehart and Winston, New York (1969), p. 364.
186 Slusher, Harold, *Bible-Science Newsletter, 13,* Jan. 1975, pp. 1-3; *ibid., 13,* Oct. 1975, pp. 1-3.
187 Clason, Clyde B., *Exploring the Distant Stars,* P. Putnam's Sons, New York (1958), pp. 220-227; Bergamini, David, *The Universe,* Time-Life Books, New York (1971), pp. 112-113.
188 Flammarion, Camille, *The Flammarion Book of Astronomy,* Simon and Schuster, New York (1964), pp. 435, 469; Batten, Alan H., *Binary and Multiple Systems of Stars,* Pergamon Press, New York (1973), pp. 222-253.
189 Hoyle, Fred, *Astronomy,* Rathbone Books Ltd., London (1962), p. 285.
190 Ostriker, J.P., and Peebles, P.J.E., *Astrophysical Journal, 186,* No. 2, Pt. 1, 1 Dec. 1973, pp. 467-480.
191 Margon, Bruce, *Mercury,* Jan./Feb., 1975, p. 6.

Section L

192 Miller, Stanley L., and Orgel, Leslie E., *op. cit.* (ref. 16), pp. 208-216.
193 Meldau, Fred John, *Why We Believe In Creation, Not Evolution,* Christian Victory Pub. Co., Denver (1972), pp. 26-51.

INDEX

FOR FURTHER STUDY

Chestnut, D. Lee. *The Atom Speaks*. San Diego: Creation-Science Research Center (hereafter referred to as CSRC), 1973.

Coppedge, James F. *Evolution: Possible or Impossible?* Grand Rapids: Zondervan, 1973.

Cummings, Violet M. *Noah's Ark: Fable Or Fact?* New York: Pyramid Publications, 1975.

Davidheiser, Bolton. *Evolution And Christian Faith.* Philadelphia: Presbyterian & Reformed, 1969.

Gish, Duane. *Evolution? The Fossils Say No!* San Diego: Creation-Life, 1973.

Himmelfarb, Gertrude. *Darwin and the Darwinian Revolution*. New York: Norton, 1968.

Klotz, John W. *Genes, Genesis and Evolution*. St. Louis: Concordia, 1970.

Kofahl, Robert E. and Kelly L. Segraves. *The Creation Explanation*. Wheaton: Harold Shaw Publishers, 1975.

Lammerts, Walter. (Ed.). *Scientific Studies in Special Creation*. Philadelphia: Presbyterian & Reformed, 1971.

Lammerts, Walter. (Ed.). *Why Not Creation?* Philadelphia: Presbyterian & Reformed, 1970.

Macbeth, Norman. *Darwin Retried*. New York: Dell, 1973.

Morris, Henry M., et al. (Eds.). *Science and Creation*. San Diego: CSRC, 1973.

Morris, Henry M., et al, (Ed.). *Scientific Creationism*. San Diego: Creation-Life, 1974.

Segraves, Kelly L. *The Great Dinosaur Mistake*. San Diego: The Beta Book Company, 1975.

Segraves, Kelly L. *The Great Flying Saucer Myth*. San Diego: The Beta Book Company, 1975.

Segraves, Kelly L. *Jesus Christ Creator*. San Diego: CSRC, 1973.

Segraves, Kelly L. *A Double Minded Man*. San Diego: The Beta Book Company, 1976.

Segraves, Kelly L. *Search For Noah's Ark*. San Diego: The Beta Book Company, 1975.

Segraves, Kelly L. *The Way It Was*. San Diego: The Beta Book Company, 1976.

Segraves, Kelly L. *When You're Dead, You're Dead*. San Diego: The Beta Book Company, 1975.

Segraves, Kelly L. (Ed.). *And God Created*, Vols. 1-4. San Diego: CSRC, 1973.

Shute, Evan. *Flaws in the Theory of Evolution*. Philadelphia: Presbyterian & Reformed, 1966.

Tinkle, William J. *Heredity*. Grand Rapids: Zondervan, 1970.

Wysong, R. L. *The Creation-Evolution Controversy*. East Lansing: Inquiry Press, 1976.

PERIODICALS

Acts and Facts, Institute for Creation Research, San Diego.

Bible-Science Newsletter, Bible-Science Association, Caldwell, Idaho.

Creation Research Society Quarterly, Creation Research Society, Ann Arbor, Mich.

CREATION-SCIENCE RESEARCH CENTER

Founded in 1970 by concerned Christians, the Creation-Science Research Center has had two principle areas of activity, the first of which has been a concerted drive to gain balanced treatment of both the evolutionary and creationist models of origins in the public schools. The Center has worked toward this goal with state and local boards of education and with local citizens groups from coast to coast. Progress has been considerable and it appears that great strides ahead will be taken in the next few years.

The Center's second area of activity has been the preparation of textbooks, popular books, and audiovisual materials which present and explain the scientific evidence which supports the Biblical record of creation and of a global flood. The aim is to see scientifically accurate creationist publications used in schools and sold in the mass markets where most people buy books. In recent years we have seen a great opening up and increase of interest in schools and among the general public. Young and old want to know both sides of the creation-evolution controversy. They are tired of the old propaganda; they want to know and to think for themselves.

The Creation-Science Research Center has had a key role in the modern movement to make known the truth about creation in order that many may come to know the Creator, Jesus Christ, who is also the Savior and Lord of creation.

For additional information write:

Creation-Science Research Center
P.O. Box 23195
San Diego, CA 92123

The Author

Robert E. Kofahl believed the gospel of Christ in 1947 as an undergraduate in chemistry at the California Institute of Technology in Pasadena, Calif., where he received the B.S. and Ph.D. degrees in 1949 and 1954. Active in campus witness and a founder of the Caltech Christian Fellowship, he soon developed an interest in the evolution-creation issue which became a long-term avocation. In 1972 after 21 years in the faculty and administration of a small Christian college, Dr. Kofahl joined the staff of the Creation-Science Research Center in San Diego, where he serves as Science Coordinator. Coauthor with C-SRC Director, Dr. Kelly Segraves, of the book, *The Creation Explanation*, and author of numerous articles and papers on evolution and creation, he has had many opportunities to lecture on this subject at schools, churches, colleges and universities.